AF751160

CONTENTS

ELSEVIER

ICRP Publication 94

Annals of the ICRP

Release of patients after therapy with unsealed radionuclides

ICRP Publication 94

Approved by the Commission in March 2004

Abstract–After some therapeutic nuclear medicine procedures with unsealed radionuclides, precautions may be needed to limit doses to other people, but this is rarely the case after diagnostic procedures. Iodine-131 results in the largest dose to medical staff, the public, caregivers, and relatives. Other radionuclides used in therapy are usually simple beta emitters (e.g. phosphorus-32, strontium-89, and yttrium-90) that pose much less risk. Dose limits apply to exposure of the public and medical staff from patients. Previously, the ICRP has recommended that a source-related dose constraint for optimisation of a few mSv/episode applies to relatives, visitors, and caregivers at home, rather than a dose limit. The present report recommends that young children and infants, as well as visitors not engaged in direct care or comforting, should be treated as members of the public (i.e. be subject to the public dose limit).

The modes of exposure to other people are: external exposure; internal exposure due to contamination; and environmental pathways. Dose to adults from patients is mainly due to external exposure. Contamination of infants and children with saliva from a patient could result in significant doses to the child's thyroid. It is important to avoid contamination of children and pregnant women. After radioiodine therapy, mothers must cease breastfeeding immediately. Many types of therapy with unsealed radionuclides are contraindicated in pregnant females. Women should not become pregnant for some time after radioisotope therapy. Technetium-99m dominates discharges to the environment from excreta of nuclear medicine patients, but its short half-life limits its importance. The second largest discharges, iodine-131, can be detected in the environment after medical uses but with no measurable environmental impact. Storing patients's urine after therapy appears to have minimal benefit. Radionuclides released into modern sewage systems are likely to result in doses to sewer workers and the public that are well below public dose limits.

The decision to hospitalise or release a patient should be determined on an individual basis. In addition to residual activity in the patient, the decision should take many other factors into account. Hospitalisation will reduce exposure to the public and relatives, but will increase exposure to hospital staff. Hospitalisation often involves a significant psychological burden as well as monetary and other costs that should be analysed and justified. Patients travelling after radioiodine therapy rarely present a hazard to other passengers if travel times are limited to a few hours.

Environmental or other radiation-detection devices are able to detect patients who have had radioiodine therapy for several weeks after treatment. Personnel operating such detectors should be specifically trained to identify and deal with nuclear medicine patients. Records of the specifics of therapy with unsealed radionuclides should be maintained at the hospital and given to the patient along with written precautionary instructions. In the case of death of a patient who has had radiotherapy with unsealed radionuclides in the last few months, special precautions may be required.

Keywords: Nuclear medicine; Radiation protection; Radioiodine; Dose constraint; Radiotherapy

ELSEVIER

ICRP Publication 94

Annals of the ICRP

Guest Editorial

RADIOLOGICAL PROTECTION AFTER NUCLEAR MEDICINE PROCEDURES

Radioactive substances have been used in medicine for over 100 years. Today, medical use of radiation is the largest, and a growing, manmade source of radiation exposure. Nuclear medicine has become an important diagnostic and therapeutic specialty, and there are nearly 100 different procedures that provide information about virtually every major organ system in the body.

The United Nations Scientific Committee on the Effects of Atomic Radiation (UNSCEAR) estimated that, worldwide, more than 30 million diagnostic nuclear medicine procedures and nearly 400,000 therapeutic procedures with radiopharmaceuticals are carried out each year (UNSCEAR, 2000). Radioactive iodine was introduced in the 1940s, and although many other radiopharmaceuticals are used in nuclear medicine, iodine-131 continues to be the most important radionuclide.

The use of unsealed radionuclides can result in exposure of other people as well as patients, and there is a need for guidance regarding the radiological protection of members of the public, relatives, and caregivers from such exposures. The International Commission on Radiological Protection (ICRP) has not provided recommendations on the criteria to follow regarding the release of patients from hospital after therapy with unsealed radionuclides, or on activity levels that require hospitalisation of the patient. Instead, the Commission has relied upon the dose limit of 1 mSv/year for the public, and the dose constraint of 5 mSv/episode for relatives, visitors, and caregivers (ICRP, 1991, 1996). These recommendations have been interpreted differently in various countries, and the dose constraint has often been inappropriately interpreted as a rigid annual dose limit.

The decision to hospitalise or release a patient should be determined on an individual basis, and should consider factors such as the residual activity in the patient, the patient's wishes, occupational and public exposures, family considerations, cost, and environmental aspects. Committee 3 established a Task Group in 1999 to review this topic. This report is one of a set of documents being developed by ICRP Committees to advise the Commission on the formulation of its next recommendations for radiological protection.

This report covers both diagnostic and therapeutic procedures, but focuses on iodine-131, which is the major source of exposure to staff and relatives from therapy with unsealed radionuclides. Precautions for the public are rarely required after diagnostic procedures, but doses to the public and relatives may need to be limited after some therapeutic procedures.

The Commission adopted the Task Group's report and its recommendations at its meeting in Vienna in April 2004.

Thyroid cancer as a result of radiation exposure appears to be a significant risk for fetuses, infants, and children. As such, particular efforts should be made to avoid exposure in these groups.

Doses to other people from patients who have received radioiodine therapy are predominantly the result of external exposure. Doses from other commonly used unsealed therapeutic radionuclides are well below public dose limits or dose constraints applied to caregivers, regardless of the radionuclide or environmental pathway considered.

This report by Committee 3 gives valuable guidance regarding whether to hospitalise or release patients. Hospitalisation will reduce exposure to the public and relatives, but will increase exposure of hospital staff and can also result in significant monetary costs that need to be analysed and justified. Patients travelling after radioiodine therapy rarely present a hazard to other passengers if travel times are limited to a few hours, and restrictions following release of patients should focus on infants and children. The Commission now recommends that the public dose limit of 1 mSv/year should apply to infants, children, and casual visitors, rather than the dose constraint of 5 mSv/episode.

Many radiopharmaceuticals can be transferred to a baby via breast milk. The Commission recommends cessation of breastfeeding, for a short period at least, after most nuclear medicine procedures, and breastfeeding should be stopped completely after a therapeutic dose of radioiodine. Otherwise, the radioiodine may induce permanent hypothyroidism or increase the risk for thyroid cancer in the infant.

With the exception of contact with a patient's urine, several studies have shown that the risk of contamination with radioiodine is generally low although not negligible. This report shows that with appropriate regulations and even without storage of urine, sewer disposal of excreta from radiotherapy patients is well within both occupational and public radiation dose limits. Storing the urine of patients treated with radioiodine appears to have minimal benefit. Radionuclides released into modern sewage systems are likely to result in doses to sewer workers or the public that are well below public dose limits.

The Commission does not explicitly state that urine should be stored or that patients should be hospitalised after therapy with high activities of radiopharmaceuticals. Instead, the Commission recommends that public dose limits and dose constraints for others should be observed. This should be followed by optimisation. With regard to release of patients following nuclear medicine therapy, optimisation and its effect on necessary behavioural restrictions may differ between individuals.

Current detection instruments are highly sensitive and can detect radiation at levels well below those that are of concern to health. Residual radioactivity may be detectable for days or weeks, and detection devices will usually detect patients for a period of weeks following radioiodine therapy. Personnel operating such detectors should be specifically trained to identify and deal with nuclear medicine patients. Many physicians give patients an information card of documentation that a medical treatment has been performed, but this may not be acceptable to security personnel.

It may be best to suggest that patients do not travel in major public areas unless they are willing to experience some inconvenience.

This report by Committee 3 clarifies the recommendations in relation to release of patients following treatment with unsealed radiopharmaceuticals. It also gives the principles for releasing patients after therapeutic administration of unsealed radionuclides.

LARS-ERIK HOLM

PREFACE

Over the years, the International Commission on Radiological Protection (ICRP), referred to below as 'the Commission', has issued many reports providing advice on radiological protection and safety in medicine. *Publication 73* (ICRP, 1996) is a general overview of this area. These reports summarise the general principles of radiation protection, and provide advice on the application of these principles to the various uses of ionising radiation in medicine and biomedical research.

Most of these reports are of a general nature, and the Commission wishes to address some specific situations where difficulties have been observed. It is desirable that reports on such problem areas be written in a style that is accessible to those who may be directly concerned in their daily work, and that every effort is made to ensure wide circulation of such reports.

A series of reports was initiated at the Commission's meeting in Oxford, UK, in September 1997. On the recommendation of ICRP Committee 3, the Commission established a number of Task Groups to produce reports on topical issues in medical radiation protection.

Some such reports have already appeared in print as *Publications 84–87* and *93*. The present report continues this series of concise and focused documents, and several more advisory reports are being prepared.

The Task Group drafting this report, on the release of patients after therapy with unsealed radionuclides, was launched at the Commission's meeting in St Petersburg, Russian Federation, in September 1999. Its initial terms of reference were to: (i) examine dose constraints for relatives and important circumstances; (ii) examine current applications and dose compared with dose rate considerations; (iii) analyse the importance of the use of conflicting approaches and cost compared with dose; (iv) consider the use of the public dose limit for dependent children with the patient as a radiation source; (v) consider recommendations concerning death, postmortem examinations, cremation, and burial; (vi) provide easily accessible recommendations for different types of therapy in different countries, taking into account that different national approaches may necessitate different practical solutions; and (vii) consider all relevant sources including rhenium, samarium, phosphorus, and yttrium. At the Commission's meeting in Bethesda, MD, USA in October 2000, these terms of reference were amended to include an expanded treatment of underlying rationale, a discussion of environmental pathways, a discussion of cost and benefit of alternative approaches, and a discussion of the ethical issues concerned.

The membership of the Task Group was as follows:

L.K. Harding (Chairman)	A. Aarkrog	D. Ash
J.-M. Cosset	S. Ebdon-Jackson	Y. Sasaki
B. Westerholm		

The corresponding members were:

K. Endo	A. Martinez	K. Parthasarathy

The membership of Committee 3 during the period of preparation of this report was:

F.A. Mettler, Jr. (Chairman)	J.-M. Cosset	C. Cousins
M.J. Guiberteau	I. Gusev	L.K. Harding (Secretary)
M. Hiraoka	J. Liniecki (Vice-Chairman)	S. Mattsson
P. Ortiz-Lopez	L.V. Pinillos-Ashton	M.M. Rehani
H. Ringertz	M. Rosenstein	C. Sharp
W. Yin		

This report aims to serve the purposes described above. In order to be as useful as possible for these purposes, its style differs in a few respects from the usual style of the Commission's publications in the *Annals of the ICRP*.

The report was approved for publication by the Commission by postal ballot in March 2004.

MAIN POINTS

- **After diagnostic nuclear medicine procedures, precautions for the public are rarely required. However, after some therapeutic procedures, doses to the public, patients's relatives, and others may need to be limited.**
- **As iodine-131 is a frequently used high-energy gamma emitter and has an 8-day physical half-life, it results in the largest dose to medical staff, the public, and relatives after procedures involving therapeutic administration of unsealed radionuclides. Other radionuclides used in therapy are primarily beta emitters (e.g. phosphorus-32, strontium-89, and yttrium-90) that pose much less risk.**
- **The major aspect of radiation therapy that needs to be controlled when releasing a patient treated with radioiodine is the external exposure of others; however, the typical doses to adults from these patients have a very low risk of cancer induction.**
- **Thyroid cancer as a result of radiation exposure appears to be a significant risk for fetuses, infants, and those under the age of 20 years. As such, particular care should be taken to avoid contamination of infants, children, and pregnant women. Internal contamination of relatives is most likely to occur 17 days after treatment. The risks from internal contamination of others are less significant than those from external exposure. Incorporation of large activities of iodine-131 that might cause hypothyroidism in an adult caregiver or relative is extremely unlikely.**
- **Very low activities of iodine-131 are observed in the environment as a result of medical uses. Even with direct release to sewage systems, the relatively short physical half-life results in doses to the public and sewer workers that are well below public dose limits and which are low compared with other sources. No environmental effects have been linked to the levels of radionuclides released as a result of medical uses of unsealed radionuclides.**
- **ICRP recommendations regarding dose limits and dose constraints related to release of patients following unsealed radionuclide therapy have been interpreted differently in various countries. These recommendations include the concept of a dose constraint of a few mSv/episode for caregivers and relatives, who should not be subject to the public dose limit. This dose constraint has often been inappropriately interpreted as a rigid annual dose limit.**
- **The ICRP now recommends that a dose constraint of a few mSv/episode should not apply to infants, young children, and casual visitors. Instead, they should be subject to the public dose limit of 1 mSv/year.**
- **Some authorities require hospitalisation of patients based on retained activity alone. Other important factors, including appropriate optimisation, are not taken into account. ICRP recommendations do not explicitly state that urine should be stored or that patients should be hospitalised after therapy with high activities of radiopharmaceuticals. Instead, the ICRP recommends that public dose limits and dose constraints for others should be observed. This should be followed by optimisation.**
- **Recent publications have indicated that assumptions and models used by some authorities to decide whether to hospitalise or release patients may overestimate actual doses to the public and caregivers.**

- **The decision to hospitalise or release a patient should be determined on an individual basis. In addition to residual activity in the patient, the decision should take many other factors into account including the patient's wishes, occupational and public exposures, family considerations, the presence of children, cost, and environmental factors.**
- **Continuation of breastfeeding is absolutely contraindicated after radioiodine therapy.**

1. INTRODUCTION

- **Nuclear medicine radiotherapy involves the use of unsealed radionuclides that have the potential to expose members of the public, relatives, and caregivers.**

(1) Radioactive substances have been used in the diagnosis and treatment of malignant and benign conditions for over 100 years. Unsealed radionuclides are radiopharmaceuticals that are injected, ingested, or inhaled, and which move through the body. These can localise in body tissues until they decay, or they can be eliminated through various pathways (such as urine). The current report deals with unsealed radionuclides, primarily those used for therapy.

(2) Nuclear medicine techniques are well established but the precautions necessary for protection of relatives, caregivers, and the general public require further guidance, particularly in light of the ICRP 1990 recommendations of a reduced annual effective dose limit to the public of 1 mSv (ICRP, 1991, 1996).

(3) Disparate interpretations of the implementation of the ICRP 1990 recommendations (ICRP, 1991) have been made, even between neighbouring countries. In addition, a number of recent scientific studies have challenged the appropriateness of assumptions that have been made in the past to estimate doses to people other than the patient.

(4) Awareness of societal costs and patient and family social issues has increased, as has awareness of environmental pathways of radionuclides and their potential effects. Regulations should be based on realistic models of radiation exposure as well as other important factors.

(5) For the purposes of this report, there are several radiation doses of interest. Organ-absorbed radiation dose is expressed in grays (Gy) or milligrays (mGy). The gray is equal to 100 rads. Occupational and public dose limits as well as dose constraints are expressed as effective dose, which is both radiation and tissue weighted. Effective dose is expressed in sieverts (Sv). The sievert is equal to 100 rem. In most medical decision-making applications, 1 Gy is equal to 1 Sv.

2. PURPOSE OF THIS REPORT

- **Regulations concerning the release of patients after administration of unsealed radionuclides vary widely around the world. Regulations should be based on assumptions that reflect the current situation accurately, or appropriate interpretations of ICRP recommendations.**

(6) The purpose of this report is to clarify aspects of radiological protection related to release of patients after administration of unsealed radiopharmaceuticals. This document also aims to promote uniform understanding and practical implementation of ICRP recommendations (particularly flexible dose constraints). It is intended for use by regulators, physicians, physicists, nurses, technologists, and appropriate administrators. These principles apply to the use of unsealed radionuclides for both diagnostic and therapeutic purposes. However, since the magnitude of exposure of other people is much less following diagnostic uses, this report focuses on exposure following therapeutic uses of unsealed radionuclides.

(7) The principles for releasing patients from hospital after therapeutic administration of unsealed radionuclides are reported here. As a basis, there is a review of underlying scientific studies, particularly with regard to how actual measurements have correlated with previous assumptions that have formed existing regulations and legislation.

(8) This report does not detail every issue related to a prescribed inpatient hospital stay, but focuses on issues related to release of the patient from a strictly controlled environment. There is extensive examination of the major source of absorbed dose to the public, relatives, and caregivers, namely external radiation. Issues regarding routes and magnitude of contamination of other people are also included.

(9) This report concentrates on the exposure pathways and doses resulting from the use of radioiodine, since this is the major source of exposure to medical staff, relatives, and caregivers from therapy with unsealed radionuclides. This report also examines radioiodine releases to the environment from medical sources, and evaluates the magnitude of doses and the potential impact in terms of risk to medical staff, relatives, and caregivers. Information on radioiodine doses to various organs and tissues can be found in *Publication 53* (ICRP, 1987).

3. TYPES AND FREQUENCY OF NUCLEAR MEDICINE PROCEDURES

- **Radioiodine treatment for hyperthyroidism and thyroid cancer is the main source of exposure to the public and relatives from patients who have received unsealed radionuclides.**

(10) Nuclear medicine is a medical specialty that involves the use of radiopharmaceuticals in the diagnosis and treatment of patients. Diagnostic procedures use short-lived gamma emitters that, when tagged to an appropriate pharmaceutical, localise in specific tissues. The images obtained give both functional and anatomical information.

(11) Common diagnostic procedures include bone scans to assess the presence of metastases, and cardiac scans to examine the perfusion of heart muscle as well as its functional capacity. UNSCEAR estimated that, worldwide, approximately 32 million diagnostic nuclear medicine procedures are performed each year (UNSCEAR, 2000). Precautions for the general public and the patient's relatives are not usually required following diagnostic procedures with radiopharmaceuticals, since most radionuclides used for diagnosis have short physical or biological half-lives. Two exceptions to this are if the patient is breastfeeding or if the patient has had an iodine-131 whole body scan to look for recurrent thyroid cancer.

(12) Compared with diagnostic applications, therapeutic treatments are much fewer in number but often use greater activities and radionuclides with longer biological and physical half-lives. Therapeutic radiopharmaceuticals are usually beta emitters, but many also have gamma emissions. Thus, they have the potential to expose other people to greater doses than diagnostic procedures.

(13) The use of radiopharmaceuticals has almost doubled over the last decade. In developed countries, UNSCEAR estimated that the frequency of radiopharmaceutical treatments increased from 0.10 per 1000 in 1985–1990 to 0.17 per 1000 in 1991–1996. UNSCEAR also estimated that, worldwide, approximately 210,000 radiopharmaceutical treatments were performed in 1985–1990 and an estimated 380,000 therapeutic procedures with unsealed radionuclides were performed between 1991 and 1996. Actual frequencies of procedures in a particular country may vary significantly from global values, and would be more useful in assessing local or regional radiation protection.

(14) The common types of therapy with unsealed radionuclides are oral or intravenous administration of liquids or capsules (systemic therapy), or instillation of colloidal suspensions into closed body cavities (intracavitary therapy). Examples of systemic therapy include sodium iodide-131 for hyperthyroidism or thyroid cancer, and strontium-89 for bone metastases. Examples of intracavitary therapy include chromic phosphate-32 for malignancies of the pleural and peritoneal cavities, and intra-articular administration for synoviectomy. Table 3.1 shows the annual estimated numbers of common therapeutic procedures with unsealed radionuclides between 1991 and 1996.

(15) The various techniques of radiopharmaceutical therapy with unsealed radionuclides are summarised briefly below.

Table 3.1. Nuclear medicine therapy: estimated annual procedures 1991–1996

Condition	Radiopharmaceutical and route	Procedures/million population in developed countries	Procedures/million population worldwide
Thyroid malignancy	^{131}I Na iodide (oral or i.v.[a])	35	15
Hyperthyroidism	^{131}I Na iodide (oral or i.v.[a])	110	42
Polycythaemia vera	^{32}P phosphate (oral or i.v.[a])	3	1
Bone metastases	^{89}Sr chloride (i.v.[a]) ^{153}Sm ethylene diamino-methylene phosphoric acid (i.v.)	5	2
Synovitis	^{90}Y colloid ^{169}Er colloid (intra-articular)	7	2
Malignant disease (other than thyroid cancer and polycythaemia vera)	^{131}I *m*-iodo-benzylguanidine (i.v.) ^{90}Y colloid (intracavitary)	[b]	[b]

Source: United Nations Scientific Committee on the Effects of Atomic Radiation (2000).
[a] Intravenous administration.
[b] Unknown.

3.1. Hyperthyroidism therapy

(16) Hyperfunction of the thyroid gland is most commonly due to an autoimmune disease (Graves's disease). Less commonly, it can be caused by a single toxic nodule (autonomous adenoma) or multinodular goitre. All of these conditions can result in overproduction of thyroid hormone with ensuing symptoms. Treatment may be with antithyroid drugs, surgery, or radioiodine. In many countries, over two-thirds of hyperthyroid patients ultimately receive radioiodine therapy. The radioiodine is concentrated by the thyroid gland, resulting in destruction of thyroid cells and a decrease in the production of thyroid hormone over a period of weeks or months. This is the most common form of nuclear medicine therapy. Treatment should be explained to the patients in a clearly written leaflet (see Appendix A).

3.2. Thyroid cancer therapy

(17) Thyroid cancer often spreads to local lymph nodes as well as the lungs and bone. Many thyroid cancers will accumulate iodine, although to a lesser extent than normal thyroid tissue. Typical therapy for many thyroid cancers is complete surgical excision of the cancer and the thyroid gland. After this, radioiodine may be given in order to destroy any residual iodine-accumulating cancer cells. Several courses of radioiodine therapy are often necessary to achieve remission or cure. This is the second most common form of therapy with unsealed radionuclides.

3.3. Bone metastases therapy

(18) Many cancers (e.g., prostate and breast) have a predilection for diffuse spread throughout the skeleton. These metastases can be very painful and, due to their

widespread nature, they are not amenable to external beam radiotherapy. Their response to chemotherapy is variable. A number of radiopharmaceuticals can be injected intravenously and will localise in the metastatic lesions to provide palliation but not cure. Common radiopharmaceuticals used are strontium-89 chloride, rhenium-186 HEDP (hydroxyethylidene diphosphonate), samarium-153 EDTMP (ethylenediaminetetramethylenephosphonate) and tin-117m DTPA (diethylenetriaminepentaacetic acid). These have to be used carefully as they may cause bone marrow depression. Due to their relatively long half-lives, they are not usually given to terminally ill patients.

3.4. Intracavitary therapy

(19) Intracavitary therapy is generally used to treat diffusely spread tumours in confined anatomical spaces, arthritis, and synovitis. Several radionuclides are used for these types of therapy. For tumour therapy, direct injection of sodium or chromic phosphate-32, gold-198 colloids or even iodine-131- or yttrium-90-labelled antibodies into confined anatomical spaces (such as the pleural space or the peritoneal cavity) is performed. For treatment of arthritis or synovitis, direct instillation of yttrium-90 FHMA (ferric hydroxide macroaggregates), dysprosium-165 FHMA, or erbidium-169 colloid into the joint space is performed. As these are beta emitters, little is needed in the way of radiation protection precautions unless the patient dies or the radionuclide leaks from the cavity.

(20) The following are much less common forms of therapy with radiopharmaceuticals.

3.5. Polycythaemia vera therapy

(21) Polycythaemia vera is a relatively rare disease that is characterised by overproduction of red and white blood cells by the bone marrow. Phosphorus-32 is given intravenously and localises in the bone. The beta emissions result in mild bone marrow suppression and reduced production of many blood elements. Since phosphorus-32 is a pure beta emitter, radiation precautions are minimal.

3.6. Intra-arterial therapy

(22) Some tumours, such as hepatomas, are highly vascularised and may not be amenable to surgery or chemotherapy. In these circumstances, it is possible to place a catheter in the arterial supply and inject insoluble radiolabelled particles that lodge in the arterioles and capillaries of the tumour and provide a local radiation dose. This is usually palliative and rarely curative. Iodine-131-labelled oil contrast and yttrium-90 glass microspheres or resin are most commonly used.

3.7. Radioimmunotherapy

(23) Radioimmunotherapy involves the use of radiolabelled antibodies that are directed against tumour-specific antigens. These agents are gaining in popularity and are currently being used for the treatment of lymphomas. The antibodies are labelled with iodine-131 or yttrium-90, and relatively large activities are injected intravenously.

4. RADIATION PROTECTION AFTER USE OF THERAPEUTIC RADIOPHARMACEUTICALS

- **Radiological protection issues following therapy with unsealed radionuclides include recommendations regarding doses to medical staff, caregivers, and the general public. After diagnostic nuclear medicine procedures, precautions for the public are rarely required. However, doses to the public, patients's relatives, and others may need to be limited after some therapeutic procedures.**
- **As iodine-131 is a frequently used high-energy gamma emitter, it is the unsealed radionuclide that results in the largest dose to medical staff, the public, caregivers, and relatives. Other radionuclides used in therapy are usually simple beta emitters (e.g., phosphorus-32, strontium-89, and yttrium-90) that pose much less risk.**

(24) Radiopharmaceuticals used for therapeutic purposes vary widely in their need for radiation protection precautions. Some radiopharmaceuticals remain relatively fixed in the body with little excretion, and primarily or exclusively emit beta rays that are essentially all absorbed within the patient's tissues. These present little risk to other people or the environment. Examples of this type of radionuclide include strontium-89, phosphorus-32, rhenium-186, yttrium-90, samarium-153, and gold-198. Patients treated with up to 200 MBq of the beta emitters (phosphorus-32, strontium-89, and yttrium-90) do not need to observe any special precautions in relation to the exposure of other people. An exception is patients receiving phosphorus-32, strontium-89, or samarium-153 lexidronam for the treatment of pain from osseous metastases. Although approximately 60–70% of these compounds will localise in the skeleton, some will be excreted in the urine (approximately 35% of the excretable fraction in the first 6 h and 80–90% in the first 48 h) post injection, and there is a need for careful hygiene to avoid contamination from urine (British Institute of Radiology, 1999; Silberstein and Taylor, 1996).

(25) Some radionuclides have multiple routes of excretion from the body before they decay, and emit gamma rays in addition to beta rays. Unless precautions are taken, this type of radiopharmaceutical can expose other people and the environment to unwanted radiation and contamination. The most common radionuclide of this type is iodine-131.

(26) The frequency of various types of nuclear medicine therapy with unsealed radionuclides is shown in Table 3.1. It is clear that treatment of thyroid conditions accounts for more than 90% of such procedures, and all of these are performed using iodine-131. Thus, for practical purposes, most issues relating to radiation protection of the public and relatives from patients who have received unsealed radionuclides for therapeutic purposes are concentrated on iodine-131.

(27) Administered activity of iodine-131 for patients treated for hyperthyroidism ranges from approximately 100 to 1000 MBq. For treatment of thyroid cancer, the administered activity ranges from approximately 4000 to 8000 MBq. The major source of radiation to members of the public and sewage treatment workers from administrations of iodine-131 will be from external radiation. For medical personnel, relatives, and caregivers, the major source of radiation is also from external

exposure, but the potential also exists for them to be exposed by contamination from the patient. Both exposure pathways should be considered when formulating recommendations or requirements.

5. CURRENT INTERNATIONAL RECOMMENDATIONS ON DOSE LIMITS AND DOSE CONSTRAINTS

- **According to previous ICRP recommendations, dose limits apply to exposure of the public and medical staff from patients. A dose constraint of a few mSv/episode (not a dose limit) applies to relatives, visitors, and caregivers at home. Dose constraints represent a source-related system of control of exposures to the individual, below which optimisation is carried out.**
- **ICRP recommendations regarding dose limits and dose constraints have been interpreted very differently in various countries. Some countries have interpreted the dose constraint as a rigid annual dose limit. ICRP recommendations do not explicitly state that patients should be hospitalised after nuclear medicine therapy.**
- **This document recommends that young children and infants, as well as visitors who are not engaged in direct care or comforting, should be treated as members of the public (i.e., be subject to the public dose limit).**

(28) In 1991, the ICRP issued recommendations concerning dose limits and dose constraints. The dose limits are summarised in Table 5.1.

(29) The ICRP 1990 recommendations state that: 'It is not appropriate to include the doses incurred by patients in the course of diagnostic examinations or therapy when considering compliance with dose limits applied to occupational or public exposure.' It is also stated that 'Medical exposure is confined to exposures incurred by individuals as part of their own diagnosis and treatment and to exposures (other than occupational) incurred knowingly and willingly by individuals in the support and comfort of patients undergoing diagnosis or treatment' (ICRP, 1991).

Table 5.1. Dose limits recommended by the ICRP[a]

Application	Occupational	Public
Effective dose	20 mSv/year averaged over defined 5-year periods[b]	1 mSv/year[c]
Annual equivalent dose in:		
Lens of the eye	150 mSv	15 mSv
Skin[d]	500 mSv	50 mSv
Hands and feet	500 mSv	

Source: International Commission on Radiological Protection (1991).

[a] Dose limits apply to the sum of the relevant doses from external exposure in the specified period and the 50-year committed dose (to 70 years of age for children) from intakes in the same period.

[b] With the further provision that the effective dose should not exceed 50 mSv in any single year. Additional restrictions apply to the occupational exposure of pregnant women.

[c] In special circumstances, a higher effective dose value could be allowed in a single year, provided that the average over 5 years does not exceed 1 mSv/year.

[d] The limit on effective dose provides sufficient protection for the skin against stochastic effects. An additional limit is needed for localised exposures in order to prevent deterministic effects.

(30) *Publication 73* (ICRP, 1996) addressed the issue of voluntary exposures in medicine, as follows:

'Friends and relations helping in the support and comfort of patients are also volunteers, but there is a direct benefit both to the patients and to those who care for them. Their exposures are defined as medical exposure but dose constraints should be established for use in defining the protection policy both for visitors to patients and for families at home when nuclear medicine patients are discharged from the hospital. Such groups may include children. The Commission has not recommended values for such constraints but a value in the region of a few millisieverts per episode is likely to be reasonable. This constraint is not to be used rigidly. For example, higher doses may well be appropriate for the parents of very sick children.'

(31) Prior recommendations on dose constraints include infants and young children who live with the patient. Since infants and young children are unable to give informed consent, are not usually involved in patient care or comforting, and are susceptible to radiation-induced thyroid cancer (especially through ingestion, see Section 6.3), it is now felt that they should be treated as members of the public, as should those visitors who are not essential to patient care or comforting. In other words, these groups should be subject to the public dose limit of 1 mSv/year.

(32) It should be emphasised that dose constraints represent a source-related system of control of exposures to the individual, below which optimisation is carried out. Optimisation is carried out after compliance with dose constraints. When a formal cost-benefit analysis has been used, optimisation is essentially a procedure of judgement that includes the magnitude of the dose, the number of people exposed, and the likelihood of incurring exposures in abnormal conditions. With regard to the release of patients who have received nuclear medicine therapy, optimisation and its effect on necessary behavioural restrictions may differ between individuals.

(33) The International Atomic Energy Agency (IAEA, 2002a) recommended actual values for dose constraints and dose limits for comforters and visitors of patients in the Basic Safety Standards (para II-9), and indicated that 'the dose limits set out in this part shall not apply to comforters of patients, i.e., to individuals knowingly exposed while voluntarily helping (other than in their employment or occupation) in the care, support, and comfort of patients undergoing medical diagnosis or treatment or to visitors of such patients. However, the dose of any comforter or visitor shall be constrained so that it is unlikely that his or her dose will exceed 5 mSv during the period of a patient's diagnostic examination or treatment. The dose to children visiting patients who have ingested radioactive materials should be similarly contained to less than 1 mSv.'

(34) The IAEA requirement refers to ingested radioactive materials and, by analogy, this should also apply to patients who receive intravenous administration of radioactive materials. Thus, the requirement is basically in line with the ICRP's recommendations, although it does not explicitly deal with avoidance of excessive dose to an individual who comforts or cares for multiple radiotherapy patients in a non-occupational setting.

6. CRITICAL PATHWAYS OF EXPOSURE FROM IODINE-131

6.1. General

- **In order of decreasing significance, the exposure to other people from patients who have received radioiodine therapy is: external exposure; internal exposure as a result of contamination; and general environmental pathways.**
- **Contamination of infants and young children with saliva from a treated patient during the first few days after radioiodine therapy could result in significant doses to the child's thyroid, and potentially raise the risk of subsequent radiation-induced thyroid cancer.**
- **Doses from other unsealed therapeutic radionuclides that are commonly used are well below public dose limits or dose constraints applied to caregivers, regardless of the radionuclide or environmental pathway considered.**

(35) Exposure of relatives, caregivers, and the public can occur in several ways:

(i) external irradiation of people close to the patient;
(ii) internal contamination of people close to the patient as a result of excreted or exhaled radioiodine; and
(iii) exposure through environmental pathways including sewage, discharges to water, incinerated sludge, or cremation of bodies.

(36) With the exception of iodine-131, other commonly used unsealed therapeutic radionuclides result in doses to the public and caregivers that are well below recommended dose limits and dose constraints, regardless of the radionuclide or environmental pathway considered. As a result, this report will focus on radioiodine.

(37) Sodium iodide-131 may enter the body through inhalation, absorption through the skin, or ingestion. It is usually administered orally in liquid or capsule form for treatment of hyperthyroidism or thyroid cancer. The iodide is rapidly absorbed from the gastrointestinal tract into the blood stream, and trapped and organified by functional thyroid tissue. Radioiodine-labelled thyroid hormone circulates on plasma-binding proteins that are metabolised by the liver and muscles. Some radioiodine is conjugated in the liver and excreted in the bile to the intestinal lumen.

(38) The salivary glands, stomach, and lactating breast contain epithelia that can maintain a concentration gradient of inorganic iodide that is approximately 1520 times the level in the plasma.

(39) Radioactive iodine is excreted primarily in the urine with smaller amounts in saliva, sweat, and faeces. A small amount is exhaled.

(40) The retained activity in the patient is a function of a number of factors including, but not limited to, the radiopharmaceutical, presence or absence of the thyroid gland, hydration, and renal function. Typical retention curves for sodium iodide-131 therapy for thyroid cancer and hyperthyroid patients are shown in Fig. 6.1.

(41) The proportions of activity of various radionuclides that are released to the sewage system are shown in Table 6.1. Driver and Packer (2001) reported on the discharge of activity measured from 174 patients undergoing radioiodine therapy for thyroid carcinoma. They found that approximately 55% of administered activity is

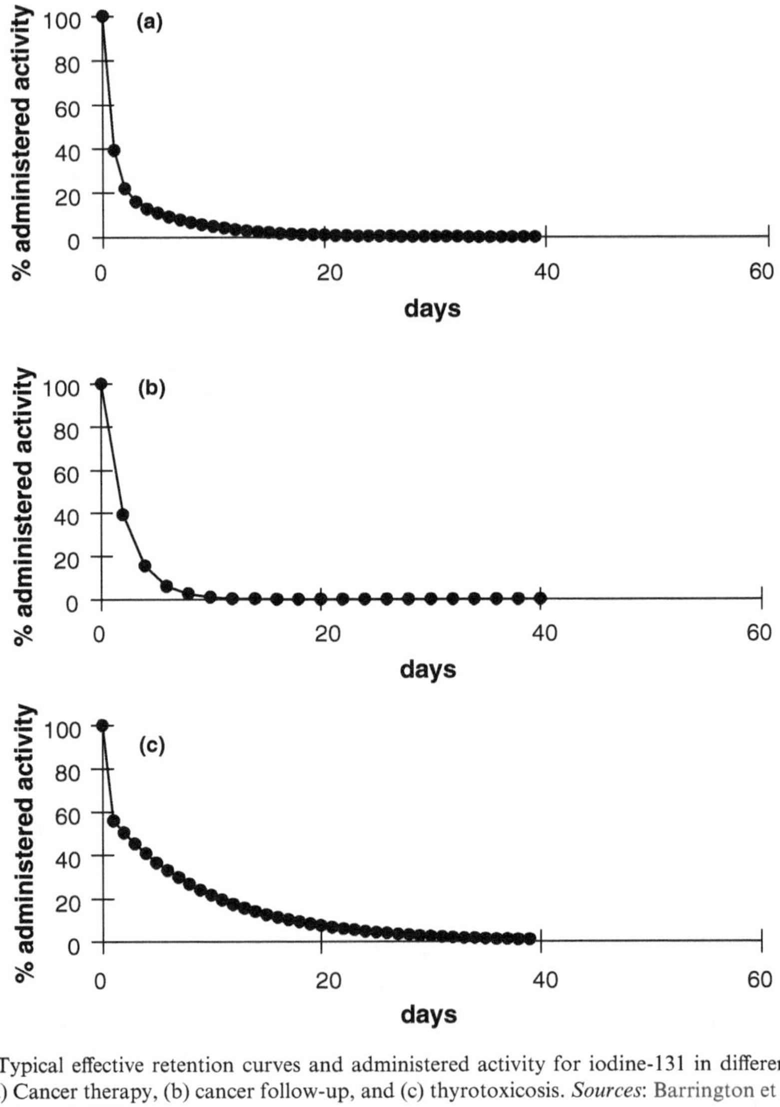

Fig. 6.1. Typical effective retention curves and administered activity for iodine-131 in different types of patient. (a) Cancer therapy, (b) cancer follow-up, and (c) thyrotoxicosis. *Sources*: Barrington et al. (1996a) and Hilditch et al. (1991).

excreted in the first 24-h period following treatment, 22% in the second 24-h period, and 6% in the third 24-h period. A total of 85% is discharged to the sewage system over the first 5 days. Historically and for regulatory purposes, it was assumed that 100% of administered activity was discharged. Use of such values is somewhat conservative and may overestimate potential environmental consequences by approximately 15%.

Table 6.1. Proportion of administered activity (until total decay) discharged to drains

Nuclide and form	For treatment of:	Discharged to sewage systems (%)
^{198}Au colloid	Malignant disease	0
^{131}Iodine	Hyperthyroidism	54
^{131}Iodine	Thyroid carcinoma	84–90
^{131}I MIBG	Phaeochromocytoma	89
^{32}P phosphate	Polycythaemia vera, etc.	42
^{89}Sr chloride	Bone metastases	92
^{90}Y colloid	Arthritic joints	0
^{90}Y antibody	Malignancy	12
^{169}Er colloid	Arthritic joints	0

Source: Thompson et al. (1994).

6.2. External dose rate from the patient

- **Doses to other people from patients who have received radioiodine therapy are predominantly due to external exposure.**

(42) For many calculations of absorbed doses from radioactive patients to other people, the activity distribution is assumed to be an unattenuated point source. External dose estimates to nearby people are usually made using the inverse square law. This is essentially true for hyperthyroid patients or thyroid cancer patients with localised metastases where the iodine is concentrated.

(43) However, if the activity is widely distributed in the patient, use of an unattenuated point-source model will overestimate doses to nearby people. A line-source attenuation correction model is more accurate and can be implemented routinely. This model would be better for patients who have received palliative treatment of osseous metastases or radioimmunotherapy (Lubin, 2002; Siegel et al., 2002a).

(44) The cumulative external exposure from a patient who has received a given activity of iodine-131 will vary by a factor of two or three depending on whether the patient is euthyroid, thyrotoxic, or being treated for thyroid cancer. As a result, a number of authors have measured dose rates from different types of patients and at different times. Culver and Dworkin (1991) measured exposure rates at various distances at times up to 11 days after administration of sodium iodide-131 for the treatment of hyperthyroidism (Table 6.2).

(45) O'Doherty et al. (1993) performed a study of dose rates in patients who received radioiodine for hyperthyroidism. The results are broadly similar and are shown in Table 6.3.

(46) Barrington et al. (1996a) measured actual external dose rates from patients who had received either ablation or follow-up iodine-131 treatment for thyroid cancer. Results are shown in Tables 6.4 and 6.5.

(47) Tables 6.4 and 6.5 show that the external dose rate per unit of administered activity decreases much more rapidly in the thyroid cancer patients. In the thyroid

Table 6.2. Measured dose rates at various times post administration and at various distances from patients treated for hyperthyroidism (μSv/h/MBq administered activity)

Distance (m)	0 days	24 days	57 days	811 days
0.6		0.059	0.034	0.024
1.0	0.046	0.022	0.014	–

Source: National Council on Radiation Protection and Measurements (1995).

Table 6.3. Mean dose rates from hyperthyroid patients at various times post administration and at various distances from patients treated for hyperthyroidism (μSv/h/MBq administered activity)

Distance (m)	Day 0	Day 1	Day 3	Day 6	Day 8	Day 10
0.1	1.3	0.4	0.3	0.2	0.2	0.1
0.5	0.2	0.1	0.1	0.07	0.05	0.04
1.0	0.06	0.05	0.04	0.03	0.02	0.02

Source: O'Doherty et al. (1993).

Table 6.4. Dose rates from ablation patients at various times post administration and at various distances from patients treated for hyperthyroidism (μSv/h/MBq administered activity)

Distance (m)	Day 0	Day 1	Day 2	Day 3	Day 4	Day 7
0.1	0.665	0.187	0.088	0.069	0.053	0.016
0.5	0.114	0.049	0.025	0.019	0.014	0.007
1.0	0.046	0.019	0.009	0.007	0.007	0.004

Source: Barrington et al. (1996a).

Table 6.5. Dose rates from follow-up patients at various times post administration and at various distances from patients treated for hyperthyroidism (μSv/h/MBq administered activity)

Distance (m)	Day 0	Day 1	Day 2	Day 3	Day 4	Day 7
0.1	0.746	0.274	0.085	0.030	0.026	0.001
0.5	0.126	0.051	0.017	0.006	0.002	0.0003
1.0	0.046	0.019	0.007	0.003	0.002	0.004

Source: Barrington et al. (1996a).

cancer patients, the thyroid has been removed so it is not retaining the radioiodine. As a result, the vast majority of administered radioiodine is eliminated in the urine in the first 2 days following treatment. Some authors have suggested that external dose measurements should be made 23 m from the patient to minimise geometric effects. The situation is different if the patient has residual foci of tumours that have accumulated a significant amount of the radiopharmaceutical.

6.3. Contamination of other people

- **Contamination of adults is much less important than controlling external exposure. However, it is very important to avoid contamination of children and pregnant women due to the sensitivity of fetal and child thyroid glands to induction of thyroid cancer. Breastfeeding must be ceased immediately by a mother following radioiodine therapy.**

(48) A common rule of thumb is to assume that no more than one-millionth of the activity being handled will become an intake to an individual working with the material. This was developed for worker intakes during normal workplace operations, accidental exposures, and public intakes from accidental airborne releases from a facility. At least two studies have shown that the same order of magnitude applies to intakes of individuals exposed to patients (Buchan and Brindle, 1970; Jacobson et al., 1978). In some countries, much more conservative values are used. For example, in the Netherlands, one-hundredth of the amount being handled is assumed for internal intake. With the exception of contact with a patient's urine, a number of studies have shown that the risk of contamination with radioiodine is generally low but not negligible. For adult relatives, the internal dose due to contamination is usually less than 10% of the external dose. Radioiodine excreted from the patient as exhaled vapour, saliva, sweat, urine, or breast milk can potentially expose relatives and caregivers. In a study where skin and thyroid doses were measured, external exposures exceeded the internal thyroid dose equivalent by a factor of over 100 (Jacobson et al., 1978).

(49) Dose conversion coefficients from ingestion and inhalation of iodine-131 by people of various ages are shown in Table 6.6. These are useful to calculate potential doses prospectively. However, measurement of activity in body fluids and in people other than the patient is more instructive.

(50) Activity in various body fluids has been measured in hyperthyroid patients treated with 200,600 MBq ofsodium iodide-131 (O'Doherty et al., 1993). Mean salivary activity collected in the first 24 h post treatment was 86.7 Bq/g/MBq administered activity (range 0.6208 Bq/g/MBq administered activity). By 3 days post treatment, this had decreased to approximately 27 Bq/g/MBq. Maximal salivary activity occurs approximately 24 h post therapy. Some authors recommend no

Table 6.6. Dose conversion coefficients (Sv/Bq) for inhalation and ingestion of radioiodine

Age group (years)	Inhalation	Ingestion
<1	7.2 E−8	1.8 E−7
1–2	7.2 E−8	1.8 E−7
2–7	3.7 E−8	1.0 E−7
7–12	1.9 E−8	5.2 E−8
12–17	1.1 E−8	3.4 E−8
Adults	7.4 E−9	2.2 E−8

Source: International Commission on Radiological Protection (1996).

mouth-to-mouth contact between the patient and their relatives for the first 48 h post therapy.

(51) The secretion of activity in palm sweat during the first 24 h post therapy was 170 Bq/cm^2 (range 131,027 Bq/cm^2), and the mean palmar secretion was 45 kBq in 24 h. There was poor correlation with administered activity or body size. Thus, the risk from iodine-131 contamination in sweat is small.

(52) Nishizawa et al. (1980) monitored iodine excretions from a number of hyperthyroid patients, with some interesting findings. In patients who received 25 mCi, activity per ml was highest in saliva (approximately 10 μCi/ml), with the exception of the first few hours post administration. It was 20-fold lower in blood (approximately 0.5 μCi/ml) and 1000-fold lower in sweat (approximately 0.01 μCi/ml). In addition, activity in saliva of these patients decreased very rapidly; at 3 days post therapy, it was approximately 1% of the initial maximum.

(53) Lassmann et al. (1998) demonstrated that in the case of radioiodine therapy, up to 0.1% of administered iodine-131 is released into the air of the therapy room. Two-thirds of relatives of patients who have been hospitalised for 2 days after radioiodine therapy will have activity in their thyroid glands that is detectable with a thyroid probe. Whether this activity is from exhalation of radioiodine or other contamination is not clear. Activity has been measured at a maximum of 4 kBq in the whole body and 0.2 kBq in the thyroid gland. This would have resulted in a thyroid dose of 2 mSv. The mean thyroid dose was 0.2 mGy. The effective dose from internal contamination of the maximally exposed relative was below the annual public dose limit of 1 mSv.

(54) Hanscheid et al. (2003) performed daily iodine-131 thyroid monitoring of personnel in a nuclear medicine therapy ward. Thyroid doses were determined and averaged to 0.35 mGy/month, showing that the incorporation risk is low provided that air exchange in the room is sufficient.

(55) Schomaecker et al. (2000) studied patients treated for hyperthyroidism and thyroid cancer. They indicated lower values than those reported by Lassmann et al. (1998), and found that the amount of radioiodine exhaled ranged from 0.008% to 0.03% of the administered activity. The percentage of radioiodine exhaled was greater in hyperthyroid patients than in thyroid cancer patients. Most of the exhaled activity was present in an organic-bound form.

(56) Wellner et al. (1998) studied the exhalation of radioiodine by hyperthyroid patients, and the resulting incorporation into their relatives. They concluded that if patients were hospitalised for 3 days, none of their relatives had effective doses exceeding 0.1 mSv. Based on their assumptions and model, if patients were treated on an ambulatory basis, the predicted effective dose to relatives would be up to 6.5 mSv and would exceed the public dose limit of 1 mSv/year.

(57) Contamination from a thyroid cancer patient is greatest at approximately 24 h post radioiodine administration, and is higher than that from hyperthyroid patients. Ibis et al. (1992) studied the contamination patterns from these patients. Removable activity from the skin was measured at 4, 24, and 48 h post administration. Activities ranged from 10 to more than 250 Bq/cm^2. There was a correlation between removable activity and administered activity. Patients who washed fre-

quently had significantly lower amounts of removable contamination. Removable contamination on surfaces that the patients touched was very variable and ranged from less than 1 to 190 Bq/cm^2. Removable activity on the rim of toilets during the first 48 h post treatment was much greater for men (approximately 1500 Bq/cm^2) than for women (approximately 20 Bq/cm^2). Salivary activity was found to be proportional to administered activity and was highest at 24 h post therapy. For patients receiving 11 GBq of activity, the 24-h activity was approximately 4 MBq/ml of saliva.

(58) Exhaled activity and mean air concentration of iodine-131 were also measured during the first 2 days post treatment. The activities in exhaled breath ranged from 20 to 190 Bq/l. The mean exhaled activity per hour during the first day post treatment was 1.5×10^{-6} Bq/h/Bq administered activity. Total exhaled activity into air for four thyroid cancer patients was 2.2–4.9 MBq. Mean room air concentrations were 0.08–0.44 Bq/l. The air in the room where the measurements were made had 190 exchanges/day. In the USA, the maximum permissible concentration for a restricted area is 0.33 Bq/l.

7. MAGNITUDE AND NATURE OF RISK FROM IODINE-131 EXPOSURE FOR RELATIVES, CAREGIVERS, AND THE PUBLIC

- **The typical doses to adults from patients treated with radioiodine have a very low risk of cancer induction. It is extremely unlikely that an adult would be contaminated with enough radioiodine to result in hypothyroidism.**
- **Thyroid cancer as a result of contamination (particularly with saliva) may be a significant risk for those under 20 years of age.**
- **Since high absorbed thyroid doses may occur in infants and young children from contamination (while remaining within a prior recommended dose constraint), and children's thyroids are very radiosensitive for carcinogenesis, the ICRP now recommends that this population should be restricted to the public dose limit of 1 mSv/year.**

(59) Dose limits and dose constraints are risk based. As such, it is useful to elucidate the nature and magnitude of the risks that are associated with these values. As mentioned earlier, the risks to others from patients who have received therapeutic amounts of unsealed radionuclides are largely derived from external exposure and, to a lesser extent, from internal contamination.

(60) Iodine-131 emits a 364-keV gamma photon and many of these emissions from the patient will expose nearby people in a predominantly uniform whole body fashion. Since the dose rates are relatively low, the risk is related to cancer induction. The risk of fatal cancer for the general population is approximately 5%/Sv (ICRP, 1991).

(61) The ICRP public dose limit is 1 mSv/year, and the dose constraint is a few mSv/episode and higher in some circumstances. Utilising a linear non-threshold approach, the risk of all fatal cancers at 1 mSv is approximately 0.005% for the general population. It is clear that these are potential risks and that increases in cancer, if present, at these dose levels are so small as to have eluded detection to date. This potential risk can be compared with the lifetime spontaneous risk of fatal cancer of 20–30%. Children are more sensitive to cancer induction than adults (by a factor of 2–3) and therefore the risk from an effective dose of 1 mSv is 0.01–0.02% for children.

(62) A Swedish study of 36,000 individuals who received diagnostic doses of radioiodine only found an increase in thyroid cancer among patients who had reported previous external radiation therapy to the neck. The thyroid dose from radioiodine among the other patients was 0.94 Gy; however, most of the patients were over the age of 20 years at exposure (Dickman et al., 2003).

(63) The hazard from internal contamination is very different in adults and children.

(64) Adults appear to be fairly resistant to induction of thyroid cancer as a result of external radiation or radioiodine (Ron et al., 1995). No dose-related increase in thyroid cancer has been found in the Chernobyl recovery workers to date (UNSCEAR, 2000). The only measurable hazard to adults in other studies occurred after high doses to the thyroid (in the range of 3 Gy), leading to subclinical hypothyroidism (Larsen and Conard, 1978). These changes only occurred after incorporation of

activities many orders of magnitude greater than those found in actual measurements of contamination from relatives of radioiodine therapy patients.

(65) Contamination of children with radioiodine is a concern in light of the increase in thyroid cancer seen after Chernobyl. Risk estimates in children after radioiodine ingestion and the Chernobyl experience are an excess absolute risk (EAR) of 1.6–2.3 per 10^4 person-year Gray (PYGy) and an excess relative risk (ERR) of 23–38/Gy (UNSCEAR, 2000). In a large study of children exposed to weapons fallout in the western USA with estimated thyroid doses of 0.46 Gy, no significant increase in thyroid cancer was found (Rallison et al., 1974). However, a later study revealed an ERR of 0.7%/mGy for thyroid neoplasms in general, but the dose–response slopes for both thyroid carcinomas and nodules were non-significant (Kerber et al., 1993).

(66) The ICRP recommendations for the public are related to effective dose alone. Except in special circumstances, children are treated as members of the public with an effective dose limit of 1 mSv/year. These are general recommendations and do not deal specifically with the issue of radioiodine and child sensitivity. It is conceivable that an older child may help in the care or comfort of a patient in a home setting if they are capable of giving consent. Given that exposure in the circumstances discussed in this report primarily concerns radioiodine and that children's thyroids appear to be sensitive to cancer induction, this issue may need further examination. With a thyroid tissue weighting factor of 0.05, it is theoretically possible that a child's thyroid could receive 20 mSv from radioiodine and still be in compliance with ICRP recommendations. The induction of thyroid cancer in children from these doses has not been shown. It is also unlikely that an older child would receive this much contamination and no external exposure. As a result, if the effective dose to a child is kept below 1 mSv/year, the thyroid dose should be substantially less than 20 mGy. This is also borne out by actual thyroid measurements made in relatives.

(67) If a nursing mother continues to breastfeed after radioiodine therapy and the child's thyroid is not ablated, the risk of thyroid cancer can be very high.

(68) One can estimate doses to a child's thyroid as a result of a specific contamination scenario in which a parent did not follow radiation protection instructions. For example, for hyperthyroid patients, salivary activity in the first day after therapy tends to average approximately 100 Bq/g saliva/MBq administered activity. If the parent received 555 MBq, the activity would be approximately 55,500 Bq/g or ml of saliva.

(69) The dose to the thyroid of an infant or young child after ingestion of iodine-131 is approximately 4.3 E − 07 Gy/Bq. As an example, if a parent did not follow precautions and an infant received 1 ml of saliva by being kissed by such a parent, the estimated thyroid dose would be approximately 2.4 E − 02 Gy or 24 mGy (ICRP, 1999). Using the preliminary data in children from Chernobyl with an ERR/Gy from radioiodine in the range of 20–30, the ERR for thyroid cancer induction would be approximately 1.0 (i.e., the natural risk would be doubled). Actual measurements from children when appropriate precautions were followed indicate lower thyroid doses (and therefore lower cancer risks) than those indicated by this example. In one study, iodine activity was detected in 25 of 89 children. In those children with

detectable activity, the thyroid dose ranged from 0.4 to 29.1 mGy (Barrington et al., 2003). This study found that some parents did not receive, understand, or follow the precautions.

(70) Doses to children from internal contamination can be compared with potential doses to a child from external exposure. If a parent who had 555 MBq of retained radioiodine activity held a child at 0.1 m for 1 h, 1 day after radioiodine administration, the external dose rate would be less (approximately 0.4 μSv/h/MBq administered or if the parent had a retained activity of 555 MBq, the external dose to the child would be 0.2 mSv). These calculations demonstrate the importance of precautions to reduce or prevent internal contamination of children and infants.

(71) Since high absorbed thyroid doses may occur in infants and young children from contamination while remaining within a prior recommended dose constraint of a few mSv/episode, and children's thyroids are highly radiosensitive for carcinogenesis, the ICRP now recommends that this population should be restricted to the public dose limit of 1 mSv/year.

8. ENVIRONMENTAL PATHWAYS OF RADIOIODINE

- **The majority of radioactivity discharged to the environment from excreta of nuclear medicine patients is from technetium-99m radiopharmaceuticals followed by iodine-131. The half-life of technetium-99m (6 h) largely limits its importance as a source of environmental exposure. Due to its half-life of 8 days, iodine-131 can be detected in the general environment after medical use. Environmental impact from these practices has not been measurable.**

(72) A number of recent publications have stressed the need for a radiation protection system that includes details of the potential impact on the environment (Copplestone et al., 2000; ICRP, 2003; Nuclear Energy Agency, 1995; Pentreath, 2002) In addition, the ICRP has a Task Group developing policy in this area at present. For the purposes of this report, the impact of released iodine-131 on the environment should be minimal. There are several reasons for this. The physical half-life of iodine-131 is relatively short (8 days), and the time it takes for the excreta of patients to be processed and returned to the ecosystem is relatively long.

(73) Movement of radioiodine in the environment has been studied for decades (Eisenbud, 1973; Ilyin et al., 1972). A vast amount of information is available as a result of radioiodine releases from accidents, nuclear weapons's tests, and intentional releases. Under these circumstances, there have been vast releases of radioiodine into the environment and food chain. Introduction into the environment has generally been via direct release into the air, although other pathways can be involved, such as sewage release into rivers and water courses that are subsequently used for irrigation.

(74) Due to its relatively short half-life, iodine-131 is not a significant contaminant in terms of soil uptake. The decay rate is rapid in relation to the growing time of crops. Radioiodine deposited on foliage is a function of deposition velocity of transfer from air to vegetation. Radioiodine is removed from foliage by weathering and other mechanisms, and the effective half-life for removal is approximately 3.5 days. Radioiodine deposited on the surfaces of plants can be ingested by cattle, and approximately 5% of ingested radioiodine appears in their milk.

(75) Regarding the release of nuclear medicine patients from hospital, the situation is very different as the radioiodine is in the body where it decays or is excreted primarily in urine, and finds its way into the environment. In the sewage management process, there is possible exposure of sewer maintenance workers and wastewater treatment operators from effluent liquids discharged to water courses and sludge. In general, the activities are very low and the dilution, dispersion, and length of time it takes to return to the food chain make this pathway of very minor importance.

(76) Radioiodine can be deposited on soil in the form of sewage sludge. Sludge may be characterised as untreated, treated, or advanced treated sludge. These terms primarily relate to processes that reduce bacteria or other pathogens, and they have little to do with removal of radioactivity other than the fact that the more processes involved and the longer it takes, the lower the released activity will be.

(77) Untreated sludge is not usually deposited on agricultural land for health reasons related to bacteria. Treated sludge may be deposited directly on crops that are serving as forage for cattle, but this is not common unless the sewage is held for up to 6 months prior to application. In some countries, there are restrictions that ban application of untreated sewage sludge on all agricultural land, and treated sludge can only be applied on grazing grassland when it is deep injected. In addition, when some of the crops are eaten raw, the time from application to harvest must be 12–30 months depending on the crop.

(78) When radioiodine is released into bodies of water, it can accumulate in a number of marine organisms. The accumulation coefficient is 200–500 for algae, 10–70 in molluscs and crustaceans, and 10–15 in the muscles of fish. In some Eastern European countries, sewage may be stored next to freshwater fish farms and put into the water to increase fish flesh production. The time from sewage treatment to 'fish feeding' is not monitored.

(79) There is at least one publication concerning the effects of radioiodine in the thyroid function of goldfish (Chavin and Cukrowski, 1968). It was found that direct intraperitoneal injections of radioiodine with activities from 0.37 kBq affected the cytological appearance of cells in the pituitary, intraperitoneal injection of 0.37 MBq (2.2 MBq/kg) achieved partial thyroidectomy, and 3.7 MBq (22 MBq/kg) caused total thyroidectomy. Atlantic salmon (LaRoche and LeBlond, 1954) and trout (LaRoche et al., 1965; Norris and Gorbman, 1965) have also been studied. The activities used to cause radiothyroidectomy were 37–185 MBq/kg. It was also noted that ablation of the thyroid gland occurs in juvenile rainbow trout placed in water with activities of approximately 1 MBq/ml for 2–3 months. Similar studies and results have also been reported in Chinook salmon, killifish, and eels (Harris, 1959; Olivereau, 1957; Olivereau and LaRoche, 1965). All reported effects occurred at activities that were many orders of magnitude greater than those from medical uses of unsealed radionuclides and discharges of patients' urine to the sewage system and subsequent effluent. However, with the use of sewage in freshwater fish farms, there may be an effect on fish thyroid glands given the 8-day half-life of radioiodine, but it is thought that this would have little impact on public health.

(80) With respect to the potential hazard, one must consider the concentration at the point of discharge, high coefficients of dilution, mixing, and significant physical decay of radioiodine along the water–algae–zooplankton–edible (by man) organism chain. The aqueous pathway by which iodine-131 may enter the human body is much less significant than the terrestrial pathway.

9. DISPOSAL OF RADIOACTIVE WASTE FROM THERAPY WITH UNSEALED RADIONUCLIDES

- **With appropriate regulations, even without storage of urine, sewer disposal of excreta from patients diagnosed or treated with unsealed radionuclides has been shown to be well within both occupational and public radiation dose limits.**
- **ICRP recommendations do not require urine to be stored. Storing the urine of patients following radioiodine therapy appears to have minimal benefit.**
- **Radionuclides released into modern sewage systems are likely to result in doses to sewer workers and the public that are well below public dose limits.**
- **Radiation detectors used at landfills to detect orphan and illicit radiation sources may detect contamination from radioiodine therapy patients or contamination in sewage sludge.**

9.1. General

(81) Public opinion regarding the disposal of radioactive waste has been studied by many authors; however, a recent study has shed some insight into public opinion regarding the source of the waste (Kelly and Finch, 2002). In general, radioactive waste from the nuclear power industry or fuel reprocessing was met with scepticism, while radioactive waste created by medical applications was acceptable given the nature of the benefits provided.

(82) A general framework for radiation protection and disposal of radioactive waste was published in *Publication 77* (ICRP, 1997). The strategies can be divided into two types: dilute and disperse, or concentrate and retain. Waste management can be a public health tool to limit public exposure from released radionuclides. It should be remembered that the primary aim of radiological protection is to provide an appropriate standard of protection for man without unduly limiting the beneficial practices giving rise to radiation exposure. The ICRP's policies are based on limiting the risk of stochastic effects by all reasonable means, but not eliminating the risk entirely.

(83) Management options for most radioactive excreta from radioiodine therapy patients can be divided into several categories similar to those above. Firstly, it can be collected, stored for decay, and then released (usually into the sewage system) at some point in the future. This is done in some countries when the patients are hospitalised. Storage of urine has not been a practical solution following patient release. A second option is to dispose of the radioactive excreta directly into the sewage system without decay. In some countries, this is done by hospitals themselves without storing urine for decay. Finally, a small amount of excreted radioactivity will be in or on materials that cannot be disposed of in the sewer. These materials may be landfilled or stored for decay (Evdokimoff et al., 1994). Each of these methods is discussed below in more detail.

(84) Direct sewer disposal is probably best for contaminated milk from a nursing mother following radioiodine treatment.

(85) Other issues have arisen as hospitals have struggled to become water efficient. This means that the facility may not be able to dilute the radioactivity to an acceptable concentration. In Canada, this is in the range of a yearly average of 200 Bq/l. As a result, at least one hospital uses multiple holding tanks (Leung and Nikolic, 1998). Depending on tank geometry, the dose rate per unit activity ranges from 0.5 to 4.0 μSv/h/GBq. These tanks are often not automatic and require proper staffing and maintenance. Such innovations may not always be necessary, and one must ensure that realistic assumptions are made regarding the environmental pathways.

9.2. Retain to decay

(86) Retaining urine for decay is a controversial subject. It is done in some countries but not in others. Ultimately the question is, what are the benefits and costs of this practice?

(87) One reason given for hospitalisation of patients is that urine can be stored and released into the sewer after decay. In many countries, patients are hospitalised to reduce the external exposure of relatives and other people, but many hospitals do not store the patients' urine. Instead, they release it directly under their authorised limits and appropriate dilution.

(88) Storing urine for decay has raised several problems. The first is determination of when the decay is sufficient (Meck, 1996). Having a requirement for 'complete' decay is unreasonable due to the nature of radioactivity. Even if one decays for 10 half-lives with iodine-131, this means storing the urine for months. Erlandsson and Mattsson (1978) recommended direct sewer disposal of all urine from radiotherapy patients without any storage.

(89) Another point is that unless there are elaborate plumbing systems in the hospital, potentially reduced public exposure will be offset by increased occupational exposure and expense. Finally, even if one hospitalises patients for 1–2 days after treatment, significant activity will be retained in the patient that will be excreted over the next few days and that will go into the sewage system from their homes.

(90) In the UK, a multidecade strategy (UK Department of the Environment, 2002) found that reduced discharge as a result of holding tanks at hospitals was not practical for most hospitals because of cost, the potential for exposing hospital staff, and because a significant proportion of discharges occur after patient release.

9.3. Sewage, sludge, and incineration

(91) At the present time, in urban areas, the predominant method of disposal of excreta from radioiodine therapy patients is via a municipal sewage system with an associated wastewater treatment plant. Contemporary sewage treatment facilities usually have primary and secondary treatment processes. The primary treatment removes grit and large floating solids, and then uses sedimentation. The secondary treatment removes unsettled solids and dissolved organic matter by flocculation or oxidation followed by sedimentation. In most cases, the effluent is then discharged

to rivers or coastal waters. The separated solids are termed 'sludge' and can be incinerated, applied to land as a fertiliser, or landfilled.

(92) In rural areas, excreta may be disposed into a local septic system or latrine. Although there are few data on this topic, disposal by a reasonably designed latrine or septic system is unlikely to result in measurable exposure of the public due to the percolation rates of water and the 8-day half-life of radioiodine.

(93) The amount of radioactive iodine in sewer water from a patient's home can be calculated from the daily urinary excretion of the patient and the volume of water released into the sewer from all uses in the house. For a single person in the UK, the latter value is approximately 1 m^3/week and for a family of four it is approximately 3 m^3/week. Approximately 55% of administered activity is excreted in urine on Day 1 post treatment, 17% on Day 2, 5% on Days 3 and 4, and 2% on Day 5. These values should not be considered alone as there will be significant dilution from other homes using the same sewage system.

(94) Activity from medical radionuclides (particularly radioiodine) in treated effluents from wastewater plants has been studied for several decades (Erlandsson and Mattsson, 1978; Prichard et al., 1981; Sodd et al., 1975). The amount of radioactivity released from sewage plants depends largely on the input, plant design, and method of ultimate disposal. The daily discharge of radioiodine from sewage plants near medical centres has been estimated to be in the range of 150–370 MBq. For many wastewater treatment plants, the effluent is sufficiently dilute that no iodine gamma spike can be detected without the use of concentration methods.

(95) Some assessments have assumed an annual occupancy rate of 200 h for workers in sewer pipes; however, in modern wastewater operations, this is more likely to be approximately 2 h/year. The estimated occupational dose to workers in a wastewater treatment facility depends not only on the episodic activity entering the plant but on the actual processing. Workers at sewage plants receive higher doses than sewer maintenance workers. The most important radionuclide contributing to the occupational dose of sewer workers is technetium-99m, whereas iodine-131 is the greatest contaminant of river water.

(96) Although releases to sewage systems are diluted by many orders of magnitude, processing at a wastewater treatment plant can result in sludge that contains easily measurable activity. The exact timing of radioiodine transit through the sewer pipes and ultimately into the sludge can be quite different due to system design and construction; however, in one Swedish study, maximal activity appeared in the sludge 2–3 weeks after radioiodine release (Erlandsson and Mattsson, 1978).

(97) The behaviour of both sodium iodide-131 and meta (^{131}I) iodobenzylguanidine (MIBG) in a municipal sewage plant have been characterised (Fenner and Martin, 1997; Martin and Fenner, 1997). This was prompted by the fact that if land application of sludge is not feasible, the sludge may be incinerated; this re-concentrates the radionuclides and the ash can trigger radioactivity detectors at landfills. Seventeen percent of administered MIBG activity appeared in the primary sludge, compared with 1.1% of administered sodium iodide-131 activity. Due to the relatively short physical half-life of iodine-131, there are few radiological concerns if

the sludge is applied to land (provided that the sludge is ploughed in or held before application to forage). Following incineration of sludge, the maximally exposed worker (after a 2-h scrubber malfunction) would receive approximately 1.7 μSv. For a typically exposed worker and the public (living 500 m from the incinerator), the committed effective dose equivalents were 1.2 and 0.06 μSv, respectively, for a 22-week incineration period.

(98) In 1998, at a meeting of the Oslo and Paris Commission, contracting parties to the 1992 Convention for the Protection of the Marine Environment of the North East Atlantic agreed to a strategy with regard to radioactive substances. The ultimate aim was to achieve concentrations in the environment close to zero for artificial radioactive substances. In the UK, a multidecade strategy was developed to comply with this goal by 2020 by working with the non-medical nuclear industry (UK Department of the Environment, 2002). It was clear that the majority of medical releases are due to short-lived technetium-99m and iodine-131, both of which decay to very low levels before entering the environment and which are therefore unlikely to have a measurable effect on the environment. The two radionuclides that originated in hospitals which were more of an issue were tritium and carbon-14. It was concluded that if medical facilities discharged at their authorised limits, a sewage treatment worker would receive approximately 0.24 mSv/year, and members of the public who ate large numbers of fish caught downstream from sewage outfalls might receive as much as 0.18 mSv/year. These are obviously upper-bound estimates based upon pessimistic assumptions. The UK Environment Agency uses a contractor to review the disposal of liquid radioactive waste to the public sewer system due to changes in sewage treatment and disposal practices, such as the increasing incineration of sewage sludge. Results from two sewage works indicated that the dose to the public was below the dose limit of 1 mSv for maximum discharge quantities. The contractor also concluded that the need for universal introduction of discharge reduction measures was not justified on the basis of current practice (Environment Agency, 2000).

(99) Crockett (2000) reported actual measurements of radioactivity at two municipal sewage plants in the UK, and used a computer model devised by the UK National Radiological Protection Board to assess behaviour of radionuclides in waste water and potential doses to sewer workers and the public. Crockett found that the majority of activity in sewage plants was from technetium-99m radiopharmaceuticals followed by iodine-131. Doses to general sewer workers and sludge press workers would be in the range of 40–80 and 150–240 μSv/year, respectively, if all hospitals, etc. were discharging at their allowable limits. However, typically, they discharged at approximately 30% of the allowable maximum limits. Public doses from release of the treated effluent would be 30–180 μSv/year for maximal allowable hospital releases, and 1–19 μSv/year for typical releases. Public doses from atmospheric releases from incinerated sludge were much less (<2 μSv/year). Crockett (2000) also pointed out that when the capacity of the sewage treatment plant was exceeded during storms, leading to untreated discharge into water, the doses to the public were not significantly affected due to the higher level of dilution by the storm water.

9.4. Landfills

(100) Recent changes in the US Nuclear Regulatory Commission (USNRC) regulations have enabled most patients requiring high-dose iodine-131 for thyroid cancer to be treated as outpatients. Before the change in regulations, most patient-contaminated articles were collected and decayed by the hospital. Now, however, these materials are often sent to landfills, following collection of waste from outpatients's homes. Many landfills now have radiation detectors to find orphan sources and illicit materials, and these detectors are quite sensitive and capable of detecting radioiodine. They are usually triggered to alarm upon detection of extremely low activity levels. This can result in an expensive search for the cause of the alarm, and in some cases, the waste has been tracked back to the patients who were billed for the expense. In some parts of the USA, the licensees are held responsible and this is why some hospitals do not release patients even though they can legally do so. It has been suggested that there is a need for reform of regulations regarding disposal of low-level rapidly decaying materials, and possibly the need for spectrometry capability.

(101) In modern landfills with leachate treatment, it is very unlikely that short-lived radionuclides used in medicine will reach groundwater before complete decay. Siegel and Sparks (2002) highlighted the paradox that released nuclear medicine patients are considered to be safe but their waste at a landfill site is considered to represent a hazard. Marcus and Aldrich (1997) made several suggestions to minimise these problems until these issues can be resolved, including: using materials that can be washed rather than paper plates or napkins; avoiding food items that cannot be eaten in their entirety (e.g., apples, barbecued ribs, etc.) in the first week after treatment because of contamination with saliva; and storing any materials that cannot be washed or disposed of in the sewer.

(102) Radiation detectors at landfills have also resulted in changes in operational policy at many hospitals. Although patients who were hospitalised for treatment with radioiodine were controlled, radioactive waste in the form of blood samples (usually contaminated with thallium-201, gallium-67 and iodine-131) and other items (such as incontinence pads) had been disposed of with medical waste. These were detected and the waste was returned to the hospitals. In one hospital, approximately 20% of surveys detected radioactivity in the waste (Evdokimoff et al., 1994). As a result, many hospitals have installed radiation detectors in the areas of the hospital through which waste bins and medical waste are transported. If radiation is detected, the source is located and is either held for further decay or disposed of with radioactive waste.

10. DECISION TO HOSPITALISE OR RELEASE PATIENTS

- **The decision to hospitalise or release a patient should be determined on an individual basis. It should not be linked solely to residual activity in the patient but should take many factors into account, including the patient's pattern of contact with other people, the patient's wishes, occupational and public exposures, family considerations, cost, and environmental factors.**
- **ICRP recommendations do not explicitly state that patients should be hospitalised after therapy with high activities of radiopharmaceuticals, but recommend that public dose limits and dose constraints for others should be observed. This should be followed by optimisation.**
- **Recent publications have indicated that assumptions used by some authorities to hospitalise patients may overestimate potential doses to the public and caregivers.**
- **Hospitalisation of patients for several days will reduce exposure to the public and relatives, but will increase occupational exposure.**
- **Isolation and hospitalisation often involve a significant psychological burden for patients and families.**
- **Hospitalisation of patients after radioiodine therapy can result in significant monetary and other costs that should be analysed and justified.**
- **Patients travelling after radioiodine therapy rarely present a hazard to other passengers if travel times are limited to a few hours.**
- **Environmental or other radiation-detection devices are usually sufficiently sensitive to detect patients who have had radioiodine therapy for a period of weeks. Personnel operating such detectors should be specifically trained to identify and deal with nuclear medicine patients.**
- **Suggested restrictions on patient behaviour vary widely in the published literature due to differences in modelling and assumptions.**
- **Actual measurements from relatives or caregivers who followed radiation protection precautions show that doses rarely approach or exceed the ICRP recommended dose constraint of a few mSv/episode.**
- **Restrictions following the release of patients should focus on the sensitive subgroup (i.e., infants and children).**

10.1. General

(103) Current recommendations regarding the release of patients after therapy with unsealed radionuclides vary widely around the world. The following have been used as criteria:

(i) ICRP dose limits and dose constraints;
(ii) dose constraints different from ICRP recommendations;
(iii) residual activity in the patient (assumed from external exposure measurements);
(iv) dose rate at a specific distance from the patient;

(v) hospitalisation because of radioiodine therapy for a specific disease (either hyperthyroidism or thyroid cancer); and
(vi) hospitalisation because of children at home.

(104) For practical purposes, it is convenient to relate activity remaining in the patient at the time of discharge to exposure to the public and relatives (Table 10.1). If this is done, it should be based on recent publications and realistic models that can be traced to dose measurements of the public. The question is whether many patients may have been unnecessarily hospitalised. This view has been expressed by authors from Russia, the UK, and the USA (Barrington et al., 1996a; deKlerk, 2000; NCRP, 1995; Shishkanov et al., 2001; Siegel, 1999). Also, Saenger and Kereiakes (1980) reported that activity-based hospitalisations may cause physicians to administer less activity than they would have liked, in order to avoid hospital stays. The use of retained activity as the sole criterion for compliance has problems as this may have little to do with subsequent patient behaviour and the ultimate dose to relatives and the public.

(105) Considerations related to hospitalisation or release of patients are shown in Table 10.2.

(106) Patients may be hospitalised after therapy with unsealed radionuclides for one of the following reasons.

Table 10.1 Activities (MBq) for release of patients depending on external dose to other people (mSv effective dose)

Radionuclide	Half-life	MBq for 5 mSv	MBq for 1 mSv
^{111}Ag	8 days	19,000	3800
^{198}Au	65 h	3500	690
^{51}Cr	28 days	4800	960
^{64}Cu	13 h	8400	1700
^{67}Cu	61 h	14,000	2900
^{67}Ga	78 h	8700	1700
^{123}I	13 h	6000	1200
^{125}I	60 days	250	50
^{131}I	8 days	1200	240
^{111}In	67 h	2400	470
^{32}P	14 days	[a]	[a]
^{186}Re	90 h	28,000	5700
^{188}Re	17 h	29,000	5800
^{47}Sc	80 h	11,000	2300
^{75}Se	120 days	89	18
^{153}Sm	47 h	26,000	5200
$^{117}Sn^{m}$	14 days	1100	210
^{89}Sr	51 days	[a]	[a]
$^{99}Tc^{m}$	6 h	28,000	5600
^{201}Tl	74 h	16,000	3100
^{90}Y	64 h	[a]	[a]
^{169}Yb	31 days	370	73

Source: US Nuclear Regulatory Commission (1997b).
[a] No value given because of minimal exposures to the public.

Table 10.2 General issues used in deciding whether to release or hospitalise patients following treatment with unsealed radionuclides

Issue	Hospitalisation	Released
Control of patient environment	High	Less
Occupational dose potential	Present	Minimal
Dose potential to relatives	Minimal	Present
Dose potential to the public	Minimal	Present
Method of disposal of waste	Sewage or storage	Sewage
Public exposure from waste	Present unless stored	Same
Monetary cost	Potentially high	Minimal
Psychological	Significant due to isolation	Minimal
Patient expiration/death	Exposure of funeral staff Possible limitation of cremation	Same

(i) Confinement and isolation of the patient will reduce the dose to the public and relatives, but will increase occupational doses to medical staff. This is an issue related to iodine-131 alone. Patients treated with pure beta emitters need not be hospitalised.

(ii) In hospital, urine can be collected and stored to reduce radioactive discharges into the sewage system. As mentioned earlier, some hospitals confine and isolate patients but do not collect and store their urine as this practice is considered to be impractical, expensive, increases staff doses, and unnecessary in light of actual measurements of discharges and their potential effect. Some authors have also suggested collection and storage of faeces, although this is generally not done as this is a minor route of excretion.

(iii) Patients with a serious illness who received therapy with unsealed radionuclides (e.g., phosphorus-32 in the peritoneal cavity of a patient with widespread metastases).

(iv) A mentally incompetent and/or incontinent patient who is incapable of following radiation safety instructions and precautions.

(v) A home situation where children would be in close contact due to physical and social constraints. If urine is not going to be collected and stored at the hospital, an alternative that is not usually discussed is for the patient to stay at a non-hospital living facility, such as a hotel, for several days. This is less expensive than staying in a hospital. This is apparently possible in some countries but not in others.

(107) Another potential hazard of hospitalisation that is rarely mentioned but which is a significant issue is the placement of a relatively well patient into an environment that may harbour antibiotic-resistant infections. A few authors have suggested that an advantage of hospitalisation is that one can calculate the iodine-131 biokinetics and evaluate absorbed doses accurately. This alone does not seem to justify hospitalisation as biokinetic studies could be performed as an outpatient procedure.

(108) In the case of children or adults who lack the capacity to consent, the physician must decide who is the most appropriate person to be given radiation protection instructions and recommendations.

(109) The ICRP cannot provide criteria for all different patients and family circumstances. It is expected that the treating physician will assess the above factors and determine what is most appropriate for each patient situation while taking radiation protection issues into account.

10.2. Occupational doses to hospital staff

(110) The major source of occupational exposure when a patient is confined to hospital is external radiation (Castronovo et al., 1986; Ho and Shearer, 1992). Barrington et al. (1996a) calculated the occupational dose to nursing staff during 7 days of care following iodine-131 administration for thyroid cancer ablation or follow-up. The results are shown in Table 10.3. Actual measurements suggest that the calculations by Barrington et al. (1996a) are very conservative. For example, Denman and Martin (2001) reported actual values in the carers of a terminally ill patient who received 800 MBq of radioiodine, and found that the nurses received a maximum effective dose of 250 μSv.

(111) Some patients must not be released after therapy with unsealed radionuclides because they are medically unstable, e.g., patients with metastatic thyroid carcinoma or hyperthyroid patients with a cardiac problem. While such patients are usually hospitalised in an isolation room, there can be emergencies that require surgery and procedures up to and including resuscitation. Griffiths et al. (2000) reported the occupational radiation doses and contamination in one such case. The patient had received 3011 MBq of sodium iodide-131 and surgery was required 6 days later. The patient ultimately died, had a postmortem examination, and was cremated. The surgeon received a dose of 20 μSv, the intensive care nurses received 40 μSv, and the mortuary assistant received 14 μSv. It was estimated that if the surgical crisis had happened immediately after administration of the iodine-131, the staff doses could have been as high as 0.8 mSv. There was a significant amount of contamination (approximately 6 MBq) in the intensive care unit, in containers of body fluids, and on laundry.

Table 10.3 Calculated cumulative dose (mSv) to nursing staff from patients for 7 days after thyroid cancer ablation treatment[a] with iodine-131

Activity (MBq)	Helpless patient	Partially helpless	Bedridden patient	Semi-ambulant patient	Ambulatory patient
1850	6.2	2.4	1.0–1.1	0.2	0.08
3700	12.6	4.8	2.1	0.4	0.16
5550	19.0	7.1	3.1	0.6	0.25
7400	25.3	9.5	4.2	0.8	0.33

Source: Barrington et al. (1996a).

[a] Values for cancer follow-up patients are within 10% of these values.

(112) Patients isolated in hospitals will contaminate surfaces that they touch, particularly the area around the toilet. These areas often need to be decontaminated before the room can be used by other patients. Iodine-131 is found in saliva, and nurses should exercise caution when dealing with vomit, and coughing or sneezing patients. Patients treated with iodine-131 will exhale some of this into room air. In at least one report, a thyroid cancer patient exhaled enough to exceed the maximum permissible concentration during the first day post treatment (Ibis et al., 1992). Occasionally, patients who have been treated with radioiodine require procedures concerning their blood (particularly haemodialysis). Following haemodialysis of a patient treated with iodine-131, no significant radioactive contamination of the dialysis equipment was reported. This has been confirmed in patients receiving other radioisotopes such as technetium-99m. After haemodialysis of a patient who was treated with iodine-131, there was slight contamination of disposable lines, waste bags, and filters that may require storage for approximately 8 weeks to allow for decay prior to disposal. The average effective half-life of iodine-131 in hyperthyroid patients on haemodialysis appears to be approximately 7 days, which is longer than that in hyperthyroid patients who are not on haemodialysis (Homer and Smith, 2002).

10.3. Psychological costs of hospitalisation

(113) Isolation in hospitals is widely recognised as a unique and depressing situation for patients, although this has rarely been taken into account with respect to patient release. Patients have indicated on multiple occasions that isolation in a hospital room suggests to them that the treatment must be much more dangerous than it actually is. The more stringent the protective measures, the more apprehensive the patients become. Patients do not realise that the major reason for isolation is to limit the cumulative occupational dose of medical staff from multiple patients and the ALARA (as low as reasonably achievable given social and economic factors) principle. Patients who are given a long list of recommended restrictions after radionuclide therapy feel that there must be a great radiation hazard. Unless the underlying reasons for isolation are explained, psychological issues are heightened (Moreno Garcia, 2001). Psychological issues should be a significant factor in the decision whether to hospitalise or release patients.

(114) When there are infants or small children in the family of a patient treated with radioiodine, there are risks from both external exposure and contamination with the patient's saliva. However, parental support of a child is very important, and considerations regarding separation of a child and parent should take the psyschological cost of such separation into account.

10.4. Cost–benefit analysis of hospitalisation

(115) As part of any radiation protection issue, cost should be considered in terms of both justification and optimisation of the practice or procedure. Very few authors or organisations have attempted to determine the costs associated with various methodologies related to release of patients after therapy with unsealed radionuclides.

Table 10.4 Estimate of the annual costs in the USA of different alternatives regarding patient release after administration of unsealed radionuclides

Alternative	Collective dose (person Sv)	Hospital retention (days)	Hospitalisation cost ($ millions)	Value of lost time ($ millions)	Record keeping ($ millions)	Psychological cost (relative)
1	184	427,000	427	256	0	High
2	298	16,000	16	10	0	Moderate
3	325	0	0	0	2.3	Low

Alternative 1 is associated with the lowest collective dose, highest cost, and highest psychological cost. If one compares this with Alternative 3, the reduction in collective dose is at a cost of approximately $2 million/person/Sv.
Source: US Nuclear Regulatory Commission (1997a).

Ideally, 'costs' should include psychological and adverse health consequences, as well as monetary costs. Cost–benefit analysis for a specific issue may vary substantially from country to country, but it does provide a tool that may help the optimisation process.

(116) The US Nuclear Regulatory Commission examined three alternatives, as follows.

(i) Alternative 1: to amend regulations to achieve an annual effective dose limit of 1 mSv for everyone other than the patient.
(ii) Alternative 2: to require confinement until the residual activity in the patient was less than 1100 MBq or the dose rate at 1 m from the patient was 0.05 mSv/h or less.
(iii) Alternative 3: to specify an effective dose limit of 5 mSv for people exposed to the patient.

(117) With the exception of a few diagnostic procedures performed using iodine-131, diagnostic procedures were unaffected by the choice of alternative, and only some therapeutic procedures were affected by choice of alternative. The results of the analysis are shown in Table 10.4. Due to the short physical half-life of iodine-131 and since the majority of exposure to others is due to external exposure, collective dose is a function of the number of people near the patient in the week or so after administration of radioiodine. For this analysis, there was an assessment of average administered activity to the patient and an estimate of the dose to the maximally exposed person other than the patient. Based upon occupancy rates, the collective dose per procedure was assumed to be approximately three times the dose to the most exposed individual. For thyroid ablation and thyroid cancer treatment, the collective dose per procedure was estimated to be 4.7 person mSv and 15 person mSv, respectively. While monetary costs differ by country, and costs are related to frequency of the procedures, Table 10.4 shows the impact of alternative choices in the USA.

10.5. Doses to others during patient travel

(118) In a recent UK study of hyperthyroid patients with less than 800 MBq of activity, the travel dose rate during their journey home averaged 49 μSv/h (range

4–252 μSv/h). Based on this and other measurements, some authors have indicated that time restrictions for private transport are not required, and restrictions will only be required occasionally for public transport (Barrington et al., 1996b; Gunesekera et al., 1996). O'Doherty et al. (1993) suggested guidelines for travel of hyperthyroid patients treated with radioiodine. These are shown in Table 10.5.

(119) Based on actual measured dose rates, Barrington et al. (1996a) calculated travel guidelines for thyroid cancer patients treated with iodine-131 for ablation or follow-up to restrict the public dose to 1 mSv/year. The results are shown in Table 10.6 and suggested restrictions are compared in Table 10.7.

Table 10.5 Suggested travel times (h) for adult hyperthyroid patients in order to restrict the public dose to 5 and 1 mSv/year[a]

Activity (MBq)	Private travel/day week 1	Private travel/day week 2	Public travel/day week 1	Public travel/day week 2
200	24 (24)	24 (24)	24 (3.5)	24 (24)
400	24 (24)	24 (24)	12 (1.5)	24 (14)
600	24 (24)	24 (24)	7 (1.0)	24 (9)
800	24 (24)	24 (24)	4 (0.5)	24 (7)

It has been assumed that private travel involves contact at 1 m with a person other than the partner, and public travel involves contact at 0.1 m with a person other than the partner.

Source: O'Doherty et al. (1993).

[a] Values in parentheses are the times suggested in order to restrict the public dose to 1 mSv/year.

Table 10.6 Travel times (h) for thyroid cancer patients in order to restrict the public dose to 1 mSv/year

Activity (MBq)	Private travel up to 24 h post dose	Private travel 24 h post dose	Private travel 48 h post dose
1850	8	20.5	24
3700	4	10	18.5, 24[a]
5550	2.5	6.5	12.5, 17[a]
7400	2	5	9, 13[a]

Source: Barrington et al. (1996a).

[a] First value is for ablation patients and second value is for follow-up patients.

Table 10.7 Comparison of suggested travel restrictions developed by different models to limit exposure to those who come in contact with iodine-131 patients to 1 mSv/year

Activity (MBq)	Private travel/day (h)	Public travel/day (h)
200	24 (24)	8.0 (3.5)
400	24 (24)	4.0 (1.5)
600	24 (24)	2.5 (1.0)
800	24 (24)	2.0 (0.5)

Values in parentheses are from O'Doherty et al. (1993).

Sources: Leslie et al. (2002) and O'Doherty et al. (1993).

10.6. Radiation detectors at borders, airports, etc.

(120) One consideration when releasing patients with internal radionuclides who have measurable gamma emissions is the unanticipated detection of such people at their place of employment, borders, airports, and other areas where there are radiation-detection systems (IAEA, 2002b). There are also considerations in releasing such patients when they return to work in areas that have radiation-detection systems (such as nuclear power plants and research laboratories). For people returning to their place of employment, there is not usually much difficulty as they are often well known and have some level of security clearance.

(121) When patients are detected at airports, etc., the situation is more difficult as these instruments are in place to restrict transport of illicit radioactive materials and detect inadvertent transport of orphan sources. Although many medical radionuclides are short lived, residual radioactivity may be detectable by current detection systems for days or weeks. Most alarms at borders are not due to illicit materials but are 'innocent' alarms due to medical radionuclides or naturally occurring radioactive materials. In the case of an alarm involving a person, the person is usually sent through the detector a second time. If the alarm recurs, the person is separated from items they are carrying and an assessment is made. Obviously it is best if the detection devices measure the gamma ray spectrum and identify the radionuclide.

(122) Many physicians give their patients an information card of documentation that a medical treatment has been performed, but this may not be acceptable to poorly trained security personnel. The IAEA have noted the possibility of illicit material being transported with legal radionuclides, and as a result, many patients will be stopped and questioned (Buettner and Surks, 2003). It may be best to suggest that patients do not do much travelling in major public areas (airports, border crossings, subways, boats, and public buildings) unless they are willing to experience some inconvenience. If such advice is provided, it should be made clear to the patient that the detection instruments are extremely sensitive and will detect radiation at levels well below those of concern to health. With current technology, it is possible to detect iodine-131 activity of approximately 0.01 MBq at 2–3 m.

10.7. Exposure in the home environment

(123) Suggested restrictions for relatives and caregivers of patients who have received radioiodine therapy vary widely. There are a number of reasons for this. If instantaneous measured dose rates at a certain distance are used as a surrogate for effective dose, the suggested restrictions will be too stringent by a factor of 2 or more. Some authors have made measurements in a phantom from either a point or a linear source, and this makes a difference. There are also assumptions in many models about the clearance kinetics of radioiodine from the patient. Finally, there are differences in assumptions regarding mode of travel, time of sleeping together, distance from relatives, etc. Many of these factors differ significantly among countries and between families in the same country. When using models and making assumptions of habits, it is probably prudent to use values that are not the mean, but which are be-

tween the 67th and the 95th percentiles. Finally, actual measured values of doses to relatives and caregivers are likely to be more valuable than models.

(124) Hilditch et al. (1991) measured mean thyroid activity as a percentage at different times after administration of sodium iodide-131 in hyperthyroid patients, and suggested that these values could be used to determine the duration of restricted contact between the public/children and the patient. In the revision of the US Nuclear Regulatory Commission regulations (1997b), release of patients after therapy was no longer based on retained activity but was based on estimated doses to the public and caregivers. The assumption was that contamination from patients was not likely to result in a significant dose to others. Johnson et al. (2002) measured the thyroid activity in relatives of thyroid cancer patients treated with radioiodine. The largest thyroid uptake resulted in an effective dose of 0.08 mSv. For a newborn, the maximal value would have resulted in an effective dose of 1.4 mSv.

(125) The hazard presented in the home by patients after radioiodine therapy has been studied over many decades. Early studies of hyperthyroid patients who were released after less than 740 MBq of sodium iodide-131 indicated that very low activities of iodine were detected in relatives, and the external radiation dose was felt to be more important but still well within the recommended dose limits at the time (Buchan and Brindle, 1970, 1971). Another early study also concluded that the external dose to relatives substantially exceeded that from cross-contamination in the majority of cases; however, detectable thyroid radioactivity was found in a number of relatives (Jacobson et al., 1978).

(126) The suggested restrictions for a patient at home vary widely in the published literature. A few examples are presented here (see Appendix B). Leslie et al. (2002) indicated that most models overestimate dose rates from radioiodine at short distances. They used adult and infant phantoms. Doses received by the adult phantom were measured at contact, 1 m, and 2 m from the patient, and doses received by the infant phantom were measured at contact in two orientations (patient cradling infant over shoulder and at waist). The doses measured in the phantoms were significantly lower than doses predicted by other models; if correct, they suggest that patient contact restrictions could be made less stringent than those currently in widespread use. Table 10.8 compares the results of O'Doherty et al. (1993) with those of Leslie et al. (2002), and reveals major differences, particularly in sleeping with a partner and absence from work. Actual measurements of doses to relatives are most realistic. The suggested guidelines of O'Doherty et al. (1993) are shown in Table 10.9.

Table 10.8 Comparison of suggested restrictions at work and home developed by different models to limit exposure to those who come in contact with iodine-131 patients to 1 mSv/year

Activity (MBq)	Absence from work (days)	Sleep apart from partner (days)	Time to restrict close contact with infant (days)
200	0 (0)	0 (15)	10 (15)
400	0 (4)	0 (20)	15 (21)
600	0 (6)	0 (24)	18 (25)
800	0 (8)	0 (26)	20 (27)

Values in parentheses are those from O'Doherty et al. (1993).
Sources: Leslie et al. (2002) and O'Doherty et al. (1993).

(127) A Belgian study of patients released after 2 days of hospitalisation following therapy for thyroid cancer found that the median dose equivalent over 2 weeks was 0.17 mSv (range 0.02–0.49 mSv) for partners with separate sleeping arrangements for 8 days, and 0.24 mSv (range 0.05–0.53 mSv) for those who slept together. For hyperthyroid patients, the corresponding figures were 1.07 mSv (range 0.22–1.27 mSv) and 1.01 mSv (range 0.05–5.23 mSv). This suggests that sleeping together may increase partner dose by 0–40% (Mathieu et al., 1997).

Table 10.9 Guidelines after iodine-131 hyperthyroid therapy to restrict dose to 5 and 1 mSv in coworkers and relatives[b]

Activity (MBq)	Absence from work (days)	Time to sleep apart[a] (days)	Time to restrict contact with child <2 years of age (days)	Time to restrict contact with child 2–5 years of age (days)	Time to restrict contact with child 5–11 years of age (days)
200	0 (0)	1 (15)	2 (15)	0 (11)	0 (5)
400	0 (3)	7 (20)	8 (21)	3 (16)	0 (11)
600	0 (6)	11 (24)	11 (24)	6 (20)	1 (14)
800	0 (8)	13 (26)	14 (27)	9 (22)	3 (16)

Source: O'Doherty et al. (1993).
[a] Assumes sleeping 1 m apart for 8 h.
[b] Values in parentheses are the guidelines suggested in order to restrict the dose to 1 mSv.

Table 10.10 Estimates of cumulative dose (mSv) to coworkers and relatives from thyroid cancer patients if no restrictions are observed

Activity (MBq)	Coworker	Partner	Child <2 years	Child 2–5 years	Child 5–11 years
1850	1, 2[a]	18, 26	25, 33	13, 17	7, 9
3700	3, 5	35, 52	50, 66	26, 35	13, 18
5550	4, 7	53, 78	75, 99	38, 52	19, 26
7400	5, 9	71, 104	100, 132	51, 69	26, 35

Source: Barrington et al. (1996a).
[a] First values are for cancer follow-up patients, and second values are for ablation patients.

Table 10.11 Suggested guidelines for thyroid cancer patients to restrict dose to 1 mSv in coworkers and relatives

Activity (MBq)	Absence from work (days)	Time to sleep apart and restrict contact with partner (days)	Time to restrict contact with child <2 years of age (days)	Time to restrict contact with child 2–5 years of age (days)	Time to restrict contact with child 5–11 years of age (days)
1850	1, 3[a]	3, 16	4, 16	3, 13	2, 10
3700	2, 7	4, 20	4, 20	4, 17	3, 13
5550	2, 10	4, 22	5, 22	4, 19	3, 16
7400	2, 12	5, 23	5, 24	4, 21	4, 17

Source: Barrington et al. (1996a).
[a] First values are for cancer follow-up patients, and second values are for ablation patients.

(128) A multicentre Belgian study reported data for relatives of 52 patients treated for hyperthyroidism with a median activity of 759 MBq and a mean activity of 370 MBq (range 185–1665 MBq) (Monsieurs et al., 1998) The mean doses to relatives of ambulatory patients and hospitalised patients were 0.6 mSv (range 0–2.0 mSv) and 0.8 mSv (range 0.4–1.7 mSv), respectively. The hospitalised patients were discharged when the dose rate at 1 m from the patient was less than 20 μSv/h. Sleeping arrangements were also studied when patients slept separately for 21 days, and all doses to partners were less than 1 mSv.

Table 10.12 Recommended restrictions on behaviour after iodine-131 hyperthyroid therapy

Restriction	30–400 MBq	400–600 MBq	600–800 MBq
All close contact with children or pregnant women (days)	9	12	14
Extended periods of contact with children or pregnant women (days)	21	25	27
Do not sleep with an adult in the same bed (days)	–	4	8
Avoid prolonged contact with other people (days)	–	–	1

Source: British Institute of Radiology (1999).

Table 10.13 Dose constraints for different categories of caregiver

Type of caregiver	Reason for dose constraint (risks, habits)	Dose constraint (mSv)
Third person	A fraction of the dose limit for the public	0.3
Relatives and close friends		
Pregnant women	Protection of the unborn child	1
Children up to 2 years old	Close physical contact with parents	1
Children aged 3–10 years	Same risk as unborn child	1
Adults up to 60 years old	2–3 times lower risk than for younger children Certain recommendations for partners Not to be applied when comforting very ill, hospitalised patients	3
Adults over 60 years old	3–10 times lower risk than for average population.	15

Source: European Commission (1998).

(129) A study of hyperthyroid patients in five UK centres with a median activity of 388 MBq (range 200–608 MBq) reported that the external doses were less than 5.3 mSv in all adults in the family (Barrington et al., 1999). The authors also studied the effect of sleeping arrangements. Two types of advice were given (A and B). With Advice A, the period of separate sleeping was 1 day for 200 MBq, 5 days for 400 MBq, 9 days for 600 MBq, and 12 days for 800 MBq. Advice B was 15 days of separate sleeping for 200 MBq, 20 days for 400 MBq, 24 days for 600 MBq, and 26 days for 800 MBq. The adult dose for those who followed Advice B was 32% of that for those who followed Advice A (see Table 10.9).

(130) Barrington et al. (1996a) published estimates of potential doses that might be received by coworkers and relatives if thyroid cancer patients disregarded the restrictions. These are shown in Table 10.10.

(131) In addition, Barrington et al. (1996a) published guidelines to restrict dose to coworkers and relatives to 1 mSv. These are shown in Table 10.11.

(132) More recent recommendations from the British Institute of Radiology (1999) indicate a somewhat shorter period of restriction, as shown in Table 10.12.

(133) In the case of a child who lacks the capacity to consent, radiation protection advice should be given to the person with parental responsibility for the child. In some countries, parents may consent on behalf of their child, but otherwise, dose limits apply. *Publication 60* (ICRP, 1991) indicated that a higher effective dose value could be allowed in a single year, in special circumstances, provided that the average over 5 years does not exceed 1 mSv/year. Prior ICRP recommendations did not mention the issue of parental consent for radiation exposure of a child. Breastfeeding should be ceased immediately following radioiodine therapy.

(134) A UK study of actual measured external doses in relatives of released hyperthyroid patients indicated that 89% of all children received <1 mSv. It is of interest, however, that 35% of children aged 3 years or younger received more than 1 mSv, indicating the need for special precautions in young children (Barrington et al., 1999). Data from a Belgian study showed that if children stayed away from home for 8 days after thyroid cancer patients were released from hospital, the dose to the children was 0.08 mSv (range 0–0.35 mSv); from hyperthyroid patients, the dose was 0.13 mSv (range 0.04–3.12 mSv) (Mathieu et al., 1999). Mathieu et al. (1999) indicated that if a dose of 1 mSv to other adults is not to be exceeded, close contact with a patient should not occur until thyroid activity is below 300 MBq. They also do not recommend close contact with young children until thyroid activity is below 100 MBq, and 50 MBq for contact with infants and pregnant women. Some authors have recommended that a short stay in hospital may be preferable for patients living with children (Reiners and Lassmann, 1999). This would avoid the scenario of a small child crawling into a sleeping mother's bed.

(135) The European Commission (1998) suggested dose constraints for different categories of caregiver. These are more detailed than those recommended by the ICRP and are shown in Table 10.13.

11. INTERNATIONAL AND NATIONAL GUIDANCE ON RELEASE CRITERIA

- **The cornerstone of release criteria are dose limits for the public and dose constraints for relatives and caregivers. In spite of this, there is wide variation in criteria used to decide whether to release or hospitalise patients. At present, the two general forms of release criteria are those based on individual situations and projected doses to other people, and those based on retained activity (usually following conservative assumptions).**
- **Since lifestyle habits differ between and even within countries, a single model for release criteria would not be appropriate optimisation. It is recommended that release of patients should be based on their family situation (rather than retained activity and the worst-case scenario). It is also recommended that when there are many contiguous countries, a uniform or similar approach to releasing patients should be developed.**

(136) The ICRP has not provided recommendations on the criteria to follow regarding the release of patients after therapy with unsealed radionuclides. Instead, the recommendations have been directed at dose limits for occupationally exposed workers in hospitals, dose limits for the public, and dose constraints for caregivers. Thus, the ICRP has not set any retained activity level to require hospitalisation. A patient may be discharged regardless of the magnitude of retained activity provided that the dose limit and dose constraint issues are met.

(137) The following text reviews the various release criteria that have been used in different countries and regions. Several different approaches have been tried. The most common approaches are release criteria based on activity or dose rate with calculations and estimates of the doses that relatives, caregivers, and the public can expect to receive. Variations in the models and assumptions used have led to widely differing release criteria for patients. As a result, some patients who live in countries with very conservative release criteria will travel to another country for therapy (so-called 'nuclear tourism').

(138) A number of recent publications have reported actual measurements of external dose rates as well as contamination potential. In light of these findings, it has been suggested that the release criteria used in some countries are overly restrictive and are focused on radiation protection without appropriate justification and optimisation. Beierwaltes and Widman (1992) indicated that while external dose rates and contamination from patients treated with radioiodine are measurable, the actual risks of detrimental effects have not been demonstrated as they are low.

(139) The International Basic Safety Standards (BSS) for radiation protection (IAEA, 1996) include the maximum activity for patients to be discharged from a hospital. The BSS indicates that '...a patient shall not be discharged from hospital before the activity of radioactive substances in the body falls below the level specified in Schedule III, Table III–VI'. The guidance level only refers to iodine-131 and is 1100 MBq, although there is a footnote that a level of 400 MBq is used in some countries as a measure of good practice.

(140) The IAEA indicated that, following radionuclide therapy, patients may not be discharged until the remaining activity subsides to an acceptable level, and that the regulatory authority should set the level according to international standards (specifically BSS, Schedule III, Table III–VI), taking local conditions and the potential exposure of other members of the patients's households into account. Additionally, activity in the patient should be estimated or measured prior to discharge, and the result should be recorded. The patients should be given written and verbal instructions of necessary precautions for protection of relatives and others with whom they may come in contact. Special precautions may be needed for the elderly or children (IAEA, 2002a).

(141) The European countries have adopted or are in the process of adopting the European BSS (European Union, 1996) and medical exposures standards (European Union, 1997). The EURATOM (1997) guidelines on radiation protection following iodine-131 therapy are quite restrictive, and can lead to 2–3 week restrictions for those with young families after 400 MBq of activity. Within the European Union, member states apply a derived residual activity constraint ranging from 95 MBq to 800 MBq, but it is set between 400 and 600 MBq in most member states.

(142) Currently, the European Thyroid Association favours treatment with iodine-131 on an outpatient basis for patients receiving up to 800 MBq of activity, provided that they abide by certain restrictions (European Thyroid Association, 1996). The EURATOM BSS 96/29 that limits the effective dose to 1 mSv/year does not apply for 'exposures of individuals, who are knowing and willingly helping other than as part of their occupation, in the support and comfort of in-patients or outpatients undergoing medical diagnosis and treatment' (European Union, 1996).

(143) The European Commission (1998) stated that 'As a general rule, treatment of thyroid cancer patients using radioactive iodine will only be performed in conjunction with hospitalisation of the patient.'

(144) The following dose constraints were proposed in a number of European publications: children and fetuses, 1 mSv; adults up to 60 years of age, 3 mSv; adults over 60 years of age, 15 mSv; and third persons/general public, 0.3 mSv (European Commission, 1998). These are identical to the regulations in Sweden. Hospitalisation is not required in Sweden if activities in the patient are less than 600 MBq of iodine-131, 1200 MBq of phosphorus-32, and 1200 MBq of yttrium-90 (Swedish Radiation Protection Institute, 2000).

(145) In Germany, essentially all patients who receive radioiodine therapy must be hospitalised for at least 48 h. The limit for discharge is 1 mSv to the public, corresponding to 3.5 μSv/h measured at 2 m. Taking an effective 7-day half-life of iodine-131 into account, the limit of activity at patient discharge amounts to 250 MBq. The activity could be higher for patients with different biokinetics provided that 1 mSv is not exceeded or for patients with a difficult psychological situation.

(146) In Japan, patients who have received iodine-131 may be released from hospital if the activity in the body is less than 500 MBq or less than 30 μSv/h at 1 m from the surface of the body. Patients who have received strontium-89 may be released if the activity remaining in the body is less than 200 MBq. Current patient release criteria in Russia are based upon not exceeding a dose of 1 mSv to the public.

(147) Several groups in Australia and New Zealand are studying radiation protection and safety issues related to patients treated with therapeutic amounts of unsealed radionuclides. These include the South Australian Hospital and University Radiation Safety Officers Group, as well as the Australian and New Zealand Society of Nuclear Medicine. The current draft discharge recommendations are essentially based on a limit of 1 mSv/year to the public, children, and pregnant women, and 5 mSv/year to consenting friends and relatives who care for the patient. The ALARA principle is used in addition to the dose limits. There is a provision that a patient should not be discharged from a controlled to an uncontrolled area if the external exposure rate exceeds 25 μGy/h at a distance of 3 m (ARPANSA, 2002). In addition, there are suggestions related to release depending on the projected excretion of activity compared with the annual limit of intake for the particular radiopharmaceutical.

(148) In the UK, an advisory committee has recently provided guidance based on national and international recommendations. This was essentially a guide to good clinical practice but it did not include any guidance on release of patients (Administration of Radioactive Substances Advisory Committee, 2000). This can be found in the Medical and Dental Guidance Notes of the Institute of Physics and Engineering in Medicine (NRPB, 2000). Patients may be discharged from hospital based on potential doses to various groups. The recommended dose constraint and dose limit per procedure for comforters and carers are 5 mSv and none, respectively. For other members of the household, the values are 1 and 5 mSv in 5 years, and for members of the general public, the values are 0.3 and 5 mSv in 5 years, respectively.

(149) In the USA, a dose-based approach to releasing patients was suggested over 30 years ago. In 1970, the National Council on Radiation Protection and Measurements stated that since the exposure rates and half-lives of various radionuclides differ greatly, a more meaningful basis for release from hospital is the possible exposure to other individuals with whom the patients are likely to associate (NCRP, 1970).

(150) In 1997, the USNRC amended its regulations for the release of patients following treatment with radioactive materials from an activity-based limit to a dose-based limit (USNRC, 1997b). The new regulations were based on the maximally exposed individual not being likely to exceed an effective dose equivalent of 5 mSv (Table 11.1). Compliance with the dose limit is demonstrated using a default table for activity or dose rate, or performing a patient-specific dose calculation. There is no specific guidance with regard to exposure of pregnant females, but it does indicate that written instructions have to be provided if a nursing child is likely to exceed an effective dose of 1 mSv.

(151) Specific instructions are required about maintaining distance from other people, minimising time in public places, precautions to reduce the spread of radioactive contamination, and the length of time that the precautions should be in effect. Instructions are generally required at 0.2 of the release values shown in Table 11.1.

(152) The activities and dose rates in Table 11.1 are conservative as they are based on physical half-life rather than effective half-life of each radionuclide. This overestimates the dose to the public as well as relatives and caregivers. With patient-specific dose calculations, the dose estimate is more realistic and appropriate, and patients can be discharged with higher activities. In fact, in the USA, competent and

Table 11.1. Activities and dose rates below which patient release is authorised by the US Nuclear Regulatory Commission

Radionuclide	Activity (GBq)	Dose rate at 1 m (mSv/h)
Au-198	3.5	0.21
Ga-67	8.7	0.18
I-123	6.0	0.26
I-131	1.2	0.07
In-111	2.4	0.2
P-32	[a]	[a]
Re-186	28	0.15
Re-188	29	0.20
Sm-153	5.2	0.06
Sr-89	[a]	
Tc-99m	28	0.58
Tl-201	16	0.19
Y-90	[a]	
Yb-169	0.37	0.02

[a] No value given because of minimal exposures to the public.

co-operative patients are now routinely discharged with activities as high as 8000 MBq of iodine-131.

(153) Coover et al. (2000) recently proposed a simplified method to conform to the USNRC regulations. By using a mathematical model, dosing charts were developed that took occupancy factors into account (Table 11.2). The results indicate that most outpatients undergoing radioiodine therapy for thyroid cancer may be treated with 7400 MBq or more. The methodology requires the physician to determine the occupancy factors (OFs) for three different periods. These are: a pre-equilibrium period of 8 h after dosing (OF_p); a constrained period of 2 days (OF_c); and an unconstrained period (OF_{uc}). The OF is determined by the percentage of time that the relative or caregiver is at a distance of 1 m from the patient. During the constrained period, the patient would sleep in a separate bedroom. An example of use of this method is shown below.

(154) A 29-year-old woman who lives with her 6-year-old daughter presents for ablation of a thyroid remnant. Her daughter is at school for 8 h/day. The mother will

Table 11.2. Maximum administered activity (MBq)[a] based upon occupancy factors for three different periods that will result in a dose of 5 mSv to relatives and caregivers from a patient being treated with radioiodine for thyroid cancer

	$OF_c = 0.125$	$OF_c = 0.125$	$OF_c = 0.25$	$OF_c = 0.25$
OF_p	$OF_{uc} = 0.25$	$OF_{uc} = 0.50$	$OF_{uc} = 0.25$	$OF_{uc} = 0.50$
0.0	20,600	12,700	15,000	10,300
0.25	13,200	9800	11,100	8300
0.75	8400	6700	7300	6000

Assumes 0.05 fractional uptake and 2-day period of constrained activity.
Source: Coover et al. (2000).
[a] Values have been rounded.

receive radioiodine on Monday morning, and she will travel home and be alone for 8 h. With counselling, the OFs for both the constrained period and the unconstrained period will be 0.25. Using the values in Table 11.2, the maximum activity that should be administered is 15,022 MBq. Since she has a child and the dose limit is 1 mSv, the value in Table 11.2 needs to be divided by 5 or 3000 MBq. Similar methodology can be applied to treatment of patients with hyperthyroidism.

(155) A regulatory analysis published soon after the appearance of the new regulations concluded that hospital stays were shorter or had been eliminated. This reduced costs significantly, and may have provided emotional benefits for the patients and relatives, and reduced occupational exposure of hospital staff. At least one author has suggested that the US approach should be adopted in other countries because 'it is of benefit for the patient while the risk to others is extremely low, not-demonstrable at dose levels of 5 mSv' (Lubin, 2002).

12. ANTIBODY THERAPY

- **Antibody therapy with radioiodine is becoming more common. Radiation protection issues are similar to those for other forms of radioiodine therapy.**

(156) To date, the literature has predominantly dealt with sodium iodide-131 for hyperthyroidism or thyroid cancer therapy. In recent years, there has been rapid spread of immunotherapy for non-Hodgkin's lymphomas. The radiotherapy works by specific binding of the immunoglobulin G, kappa monoclonal antibody ibritumomab to the C20 antigen found on the surface of malignant and normal B lymphocytes. The antibodies can be labelled with iodine-131 or, more commonly, with yttrium-90.

(157) In cases where iodine-131 labelling is used, the thyroid is usually blocked with saturated potassium iodine. Any iodine-131 that dissociates is excreted rapidly in the urine. Urinary excretion is approximately 7% during the first week. From the start of therapy for a period of 1 week, it is recommended that a condom should be used during intercourse, and deep kissing should be avoided, as should other forms of transfer of body fluids. Patients are advised to wash their hands thoroughly after using the toilet. When the antibody is labelled with yttrium-90, radiation protection issues are minor as yttrium-90 is a pure beta emitter with a beta path length of 5 mm (100–200 cell diameters).

(158) Recent research on dose rates in antibody therapy patients has indicated that for mean administered activities of 3100 MBq, the mean measured dose rate at 1 m immediately after administration is approximately 0.11 mSv/h. This is approximately 60% of the calculated dose rate using a point-source model. The mean calculated dose to a maximally exposed person nearby (0.25 occupancy rate at 1 m) is approximately 3 mSv (Siegel et al., 2002b). Actual measurement of the doses to relatives of patients who had received anti B1 immunotherapy with activities ranging from 0.94 to 4.77 GBq yielded values ranging from 0.17 to 4.09 mSv (Rutar et al., 2001). It is of interest to note that the measured doses were approximately one-third of the doses calculated using US Nuclear Regulatory Commission methodology.

13. OTHER ISSUES

- **Records of the specifics of therapy with unsealed radionuclides should be maintained at the hospital and given to the patient along with written precautionary instructions.**
- **In the case of death of a patient who had received radiotherapy with unsealed radionuclides in the last few months, it is advisable to contact a radiation protection specialist or the hospital that treated the patient to determine what precautions are necessary in accordance with national regulations.**
- **Many types of therapy with unsealed radionuclides are contraindicated in pregnant females.**
- **Women should not become pregnant for a variable period of time after radioisotope therapy. The time depends on the type of therapy, and is a function of clearing the radionuclide from the body and ensuring that the underlying disease is controlled.**

13.1. Records

(159) On release from hospital after therapy with unsealed radionuclides, all patients should be given written details (such as a wallet-sized card) of the radionuclide, the physical or chemical form, the administered activity, and the name and telephone number of the treating physician. In addition, patients should be given written radiation safety precautions and information about when these precautions can be terminated. An example of such a card is shown in Appendix C.

13.2. Death, postmortem examinations, burial, and cremation

(160) While it is common practice to issue written information indicating that a patient has received radionuclide therapy, one cannot always assume that the patient will have it with them in case of an emergency or death.

(161) Actual measurements from a postmortem examination of a patient who had received 800 MBq of radioiodine were reported by Denman and Martin (2001). The patient, who was terminally ill, died 3 days after administration of the radioiodine, and the postmortem examination was delayed for 2 weeks. The pathologist was estimated to have received a maximum whole body dose of 400 μSv. Contamination measurements were: pathologist's hands, 5 Bq/cm^2; towels, 1.8 Bq/cm^2; saw, 5 Bq/cm^2; instruments, 0.5 Bq/cm^2; plastic sheet, 0.8 Bq/cm^2; scales, 0.4 Bq/cm^2; and floors and walls, 1.1 Bq/cm^2.

(162) There are at least two reports of postmortem examinations of patients who had received radioiodine therapy for thyroid cancer. One patient had received 1850 MBq of radioiodine for metastatic thyroid cancer (Parthasarathy et al., 1982). The patient died 7 days later having excreted very little radioiodine in the urine. The radiation dose level after death ranged from 0.1 to 0.5 mGy/h at 10 cm from the body surface. Several pathologists performed the postmortem examination in order to

keep individual doses low. The most-exposed pathologist received a whole body dose of 220 μSv and a hand dose of 5.5 mSv.

(163) The second report concerned the postmortem examination of a patient who had received 7.4 GBq of iodine-131. The patient died 10 days after treatment, with 1.85 GBq remaining in the cadaver (Johnston et al., 1979). Most excretion curves for such patients would have predicted less remaining activity. Poor hydration or renal function may have precluded normal excretion rates. The measured dose rate at the surface of the chest was approximately 0.6 mSv/h. With appropriate radiation protection precautions and a maximum time limit of 90 min, the pathologist had a measured dose of approximately 0.2 mGy in his pocket ionisation chamber and about three times this on his hands.

(164) There is little guidance regarding the safe handling of corpses. In Australia, the Code of Practice (National Health and Medical Research Council, 1987) includes the following information: 'Radioactive material administered by oral, intravenous, or intracavitary routes may become distributed throughout the body or may be selectively held in specific organs. The degree of possible hazard presented by the patient reduces with the lapse of time from the time of administration of the radioactive material. This reduction arises partly from the normal decay of the radioactivity of the material used and partly from the excretion of the material from the body. A patient who had received very small amounts of radioactive materials used for diagnostic or tracer tests presents no hazard to anyone called upon to attend the corpse and is not considered further.' With regard to autopsies, the Code states that 'if a corpse contains less than any of the following activities:

150 MBq radon-222, colloidal yttrium-90, or gold-198;
300 MBq phosphorus-32;
450 MBq iodine-131, sealed yttrium-90, or gold-198;

the procedures normally observed during autopsy are adequate for the examination unless such examinations are carried out frequently in the same institution. If activities are greater than that quoted above or if autopsies of the above activities are carried out frequently in the same institution, the pathologist is advised to consult with the radiation safety officer'. In addition, the Australian Code of Practice (National Health and Medical Research Council, 1987) indicates that if the radioactive material used for treatment has been selectively absorbed in a particular organ (e.g., iodine-131 in the thyroid), the organ should be excised before the examination proceeds and removed from the work area. It may later be disposed of with the body. If the radioactive material is distributed within certain body fluids, the fluids should be drained off, using suitable equipment, before the examination proceeds and the fluids may be disposed of safely via the sewage system with regard to legislation requirements.

(165) For burial of the corpse, the Australian Code of Practice (National Health and Medical Research Council, 1987) states that: 'No special precautions are normally required in direct burial, without embalming. No special precautions are required for embalming if activities do not exceed those levels mentioned above. If

the activities are greater then the corpses should not normally be embalmed but if embalming is required, a radiation safety officer should be consulted.'

(166) For cremation, the Australian Code of Practice (National Health and Medical Research Council, 1987) states that no special precautions are required if the corpse does not contain more than 1000 MBq of yttrium-90, iodine-131, gold-198, iodine-125, or radon-222, or 400 MBq of phosphorus-32. Corpses containing in excess of these levels should be stored until these limits are reached.

(167) The UK regulations for cremation and/or burial after iodine-131 therapy state that activity should not be more than 400 MBq (NRPB, 1988). More recent guidance has confirmed the 400 MBq value but states that if the activity is greater than 400 MBq, it may still be possible to bury or cremate the body but advice should be sought from the radiation protection advisor (NRPB, 2000).

(168) Swedish regulations are very similar to those in Australia. The Swedish regulations indicate that the highest activities at which a postmortem examination can be performed without radiation protection measurements are 600 MBq of iodine-131, 400 MBq of phosphorus-32, and 200 MBq of yttrium-90. For cremation without radiation protection measures, the values must not exceed 1200 MBq of iodine-131, 400 MBq of phosphorus-32, and 1200 MBq of yttrium-90 (Swedish Radiation Protection Institute, 2000).

(169) In some countries, such as Japan, cremation is much more common than burial. As a result, there is greater concern about the potential release of radionuclides to the environment. UNSCEAR (2000) reported that there were 0.0073 iodine-131 thyroid cancer treatments per 1000 population and 0.023 iodine-131 hyperthyroidism treatments per 1000 population between 1991 and 1996. Assuming that the population at that time was approximately 110 million, approximately 2300 hyperthyroid and 730 cancer patients would be treated annually. It seems unlikely that more than 1% of these patients (approximately 30 patients) would die within a few weeks of treatment, so the amount of radioiodine potentially released to the public would be very small.

(170) Cremation of corpses containing bone-seeking radionuclides used for palliation of osseous metastases is more of a problem since the radionuclides have a relatively long half-life (Aerts, 2000). Typically, this therapy is not used if life expectancy is less than 3 months; however, if the patient dies before this time, it can take up to 1 year for the activity of strontium-89 to decay to 1 MBq. Storage of the corpse is impractical given the physical half-life of 50.5 days. A study of several cremations in Australia indicated that most, if not all, of the activity remains with the bone dust. Depending on the family's intention for the ashes, storage may be needed in order to comply with local regulations. In the USA, there is no problem with cremation if bodies contain less than 74 MBq for all radionuclides, except iodine-131 which has a limit of 7400 MBq/year for a particular crematorium (Silberstein et al., 2003).

(171) There was an interesting anecdotal discussion recently concerning the demand of an undertaker. It was pointed out that a coffin lined with 2.6 mm of lead would weigh over 1000 kg and this would be impractical from a number of respects (Osborn et al., 2002). There is at least one report in which a patient who had received

radioiodine therapy for thyroid cancer died and was placed in a casket lined with 1.6 mm of sheet steel. The dose rate measured at the chest surface of the cadaver was 0.6 mSv/h, and after placement in the casket, the dose rate was 0.5 mSv/h at the surface of the casket, indicating little shielding effect.

13.3. Breastfeeding

(172) Many laboratories ask all females to indicate if they are breastfeeding, as many radiopharmaceuticals can be transferred to a baby via breast milk. Cessation of breastfeeding, for a short period at least, is recommended for most nuclear medicine studies. Breastfeeding should be ceased completely after a therapeutic dose of radioiodine. If this is not done, the infant may become permanently hypothyroid or be at high risk for subsequent thyroid cancer. Doses to infants as a result of breastfeeding are shown in Table 13.1.

(173) It may be beneficial for women to cease breastfeeding 2–3 weeks before receiving radioiodine therapy. The advantages are that the breasts will stop producing milk, dose to the breast tissue will be reduced, there is no danger of non-compliance, and contaminated brassieres and breast binders will not be an issue.

13.4. Pregnant females

(174) In 2000, the ICRP published a document on issues related to pregnancy and occupational radiation, and indicated that 'dose limits for the fetus are broadly comparable with those for the general public' (ICRP, 2000). It was also stated that after pregnancy was declared, the dose to the conceptus should not exceed approximately 1 mGy for the rest of the pregnancy.

(175) As certain radiopharmaceuticals, including iodine-131 and phosphorus-32, can cross the placenta rapidly, the possibility of pregnancy should be very carefully considered before such radionuclides are given for therapy or for a whole body iodine-131 scan for thyroid carcinoma. As a rule, a pregnant woman should not be

Table 13.1. Dose (Sv/Bq) to an infant as a result of breastfeeding from a mother after a single ingestion of iodine-131

Time of ingestion	Effective dose
26 weeks before pregnancy	0
5 weeks of pregnancy	0
15 weeks of pregnancy	3.4E – 16
35 weeks of pregnancy	1.3E – 10
Birth + 1 day	5.4E – 08
Birth + 10 days	5.4E – 08
Birth + 20 days	5.4E – 08

Source: Draft report of ICRP Committee 2 Task Group on doses to the infant from ingested radionuclides in mothers' milk.

treated with a radioactive substance unless the radionuclide therapy is required to save her life. In that extremely rare event, the potential absorbed dose and risk to the fetus should be estimated and conveyed to the patient and the referring physician. Considerations may include terminating the pregnancy.

(176) In women, thyroid carcinoma comprises over 80% of cancers of the head and neck between the ages of 15 and 45 years. Thyroid cancers are relatively non-aggressive compared with most other cancers. As a result, both surgery and radioiodine treatment are often delayed until after pregnancy. In general, if any therapy is necessary during pregnancy, it will be performed during the second or third trimesters.

(177) Radioiodine can cross the placenta easily and the fetal thyroid begins to accumulate iodine at approximately 10 weeks' gestational age. Radioiodine therapy is essentially contraindicated in patients who are known to be pregnant. If radioiodine treatment of thyroid carcinoma is necessary, it should be delayed until after delivery. If this is done, the physician should be aware that radioiodine is excreted in breast milk, and breastfeeding should be ceased completely after a therapeutic dose.

(178) A major problem occurs when a female, who is not thought to be pregnant, is treated for thyroid carcinoma and is found out to be pregnant after administration of the radioiodine. Menstrual history is often not adequate to ensure that a patient is not pregnant. In most developed countries, it is common practice to obtain a pregnancy test prior to high-dose iodine-131 scanning or therapy for women of child-bearing age unless there is a clear history of prior tubal ligation or hysterectomy. Despite the above, pregnant women do still receive treatment, either because of false histories or because the pregnancy is at such an early stage that the pregnancy test is not yet positive.

(179) Most commonly, the pregnancy is early and the major problem is fetal whole body dose due to gamma emissions from radioiodine in the maternal bladder. During pregnancy, the whole body dose to the conceptus is in the range of 50–100 μGy/MBq of administered activity. Dose to the embryo and fetus from maternal ingestion of radioiodine-131 is shown in Table 13.2.

(180) The dose can be reduced by oral hydration of the patient and by encouraging frequent voiding. This is a common recommendation for all patients, whether pregnant or not.

(181) If the conceptus is more than 8 weeks' gestational age (and the fetal thyroid can accumulate iodine) and the pregnancy is discovered within 12 h of iodine administration, a maternal dose of 60–130 mg stable potassium iodide will partially block the fetal thyroid and reduce the thyroid dose. More than 12 h after radioiodine administration, this intervention is not very effective.

(182) Maternal hyperthyroidism can occur during pregnancy. The diagnosis can be made on the basis of serum hormone determinations rather than radioiodine uptake studies or thyroid scintigraphy. Treatment can often be delayed until after pregnancy, and the patient can be treated with drugs in the interim. Again, the major problem is discovering that a patient is pregnant after they have received a therapeutic dose of radioiodine.

Table 13.2. Dose coefficients (Sv/Bq) for the offspring from acute maternal ingestion of iodine-131 during and after pregnancy

Time (weeks)[a]	Most exposed organ	Highest organ dose	Effective in-utero dose	Effective postnatal dose	Total effective dose to offspring
−130	–	<1E−15	<1E−15	<1E−15	<1E−15
−26	–	<1E−15	<1E−15	<1E−15	<1E−15
At conception	All	7.8E−11	7.8E−11	<1E−15	7.8E−11
5	Thyroid	2.4E−11	8.1E−11	<1E−15	8.1E−11
10	Thyroid	3.2E−09	2.1E−10	<1E−15	2.1E−10
15	Thyroid	2.4E−07	1.2E−08	4.3E−15	1.2E−08
25	Thyroid	6.8E−07	3.4E−08	3.3E−12	3.4E−08
35	Thyroid	1.1E−06	5.5E−08	5.3E−09	6.0E−08

Source: International Commission on Radiological Protection (2002).
[a] Intake at the indicated time (weeks); negative times are prior to pregnancy.

Table 13.3. Periods for avoiding pregnancy after radioisotope therapy to ensure that the dose to the fetus will not exceed 1 mGy

Nuclide and form	For treatment of:	All activities up to: (MBq)	Avoid pregnancy (months)
^{98}Au colloid	Malignant disease	10,000	2
^{131}I Na iodide	Hyperthyroidism	800	4
^{131}I Na iodide	Thyroid cancer	6000	4
^{131}I-MIBG	Phaeochromocytoma	7500	3
^{32}P phosphate	Polycythaemia vera etc.	200	3
^{89}Sr chloride	Bone metastases	150	24
^{90}Y colloid	Arthritic joints	400	0
^{90}Y colloid	Malignancy	4000	1
^{169}Er colloid	Arthritic joints	400	0

Source: Administration of Radioactive Substances Advisory Committee (2000).

(183) Patients treated with radioiodine can be an external radiation source to pregnant relatives. Perhaps more importantly, these patients must be careful not to transfer radioiodine contamination to pregnant relatives by direct contact or through indirect means.

13.5. Subsequent pregnancy after radionuclide therapy

(184) Occasionally, the question arises about the advisability of becoming pregnant after a nuclear medicine examination or treatment. Most female patients are advised not to become pregnant for at least 6 months after therapy with radioiodine. This is not primarily based upon potential heritable radiation effects or radiation protection considerations per se, but is based upon the need to be sure that: (1) the hyperthyroidism or cancer is controlled; and (2) another treatment with radioiodine will not be needed when the patient is pregnant.

(185) Phosphorus-32, strontium-89, and MIBG are occasionally used for therapy. In order to keep the dose to the fetus below 1 mGy, pregnancy should be avoided for 3, 24, and 3 months, respectively. The ICRP has recommended that a woman should not become pregnant until the potential fetal dose from remaining radionuclide does not exceed 1 mGy. This is not usually a consideration except for radioiodine therapy or radiopharmaceuticals labelled with iron-59 (for metabolism studies) or selenium-75 (for adrenal imaging). As a result of the long physical half-lives of these radionuclides and their long residence times in the body, it is recommended that pregnancy should be avoided for 6 and 12 months, respectively. Advice in the UK is presented in Table 13.3 (Administration of Radioactive Substances Advisory Committee, 2000).

APPENDIX A. SAMPLE PATIENT INFORMATION SHEET FOR HYPERTHYROIDISM

Radioiodine treatment for hyperthyroidism

Your questions answered

Why do I need treatment?
You have a condition called hyperthyroidism. This means that your thyroid gland is overactive. If it is not properly treated, your health may be affected in the future.

What is radioiodine treatment?
Radioiodine treatment uses a form of iodine that is radioactive. Iodine is taken up by the thyroid gland, so only a small amount of radioactivity is needed. Your doctor considers that this is the best form of treatment for you.

Where does the radioactivity go?
Most of the iodine is taken up by the thyroid. The rest of the iodine mainly passes out of your body in urine.

How is the iodine given?
Radioiodine is colourless and tasteless. You will be asked to swallow a liquid or a capsule containing the radioiodine.

Will I have any side-effects?
Minor side-effects (such as a sore throat) may be associated with treatment.

What about my medication?
Your hospital doctor may have given you instructions about the medication you are taking.

Is radioiodine treatment safe?
Radioiodine has been used for over 40 years to treat hyperthyroidism. Patients treated this way have been studied carefully. This form of treatment is considered to be safe and effective.

Are there any extra risks in having children afterwards?
There has been no effect on the health of children of patients who have had radioiodine. However, we do ask you to avoid pregnancy and breastfeeding for several months after radioiodine treatment, just in case you need another radioiodine treatment.

Is there a risk to others?
No, not if you follow the radiation precautions given to you by your physician. You will be given some simple precautions to follow when you attend for your treatment. These are merely to avoid any unnecessary radiation to others. Others may need to be made aware of the risk of exposure.

Will I need to see a doctor after the radioiodine treatment?
You should be seen by your doctor after the treatment and have blood tests taken. These are to check how your thyroid gland has responded. It takes 2–3 months for radioiodine to take effect.

How many radioiodine treatments will I need?

Occasionally, a second or even a third treatment is necessary. The blood tests after your first treatment will show whether further treatment is needed.

Are there any long-term effects?

Radioiodine is a very safe treatment. However, your thyroid gland will probably be underactive after your treatment. This could happen within a few months or may take many years. That is why the blood tests to check the function of your thyroid are important and should be performed regularly for the rest of your life. If your thyroid becomes underactive, you will be started on thyroxine treatment. This has no side-effects and only needs to be taken once a day.

We want you to understand what the treatment involves. If you have any other questions, please ask when you come to the hospital for treatment when you will have the opportunity to discuss your concerns with us.

APPENDIX B. SAMPLE INSTRUCTIONS FOR RADIATION PROTECTION AFTER THERAPEUTIC ADMINISTRATION OF RADIOIODINE

To be followed for approximately 1 week
Source: European Commission (1998).

Immediately
Do not eat for 1 h after oral administration of radioiodine. If you vomit within 4 h, try to do so in a waste bin and inform the nuclear medicine department immediately.

Travel
Avoid public transport if possible. If necessary, travel by public transport should be restricted to about 2 h. Do not take a long trip (6 h or more) with relatives or caregivers. Try to sit at least 1 m from others.

At home
Avoid prolonged physical contact. Stay as far away as possible from everyone at home, more than 1 m at all times and more than 2 m for extended periods of time. Do not remain within 1 m of any individual for more than 6 h/day.

Sleep in a separate bed and in a separate bedroom if possible.

Drink plenty of liquids.

Do not share food or drinks with others. Some recommend using disposable dishes and utensils but this is not necessary and may cause difficulties at landfills. It is acceptable to simply wash the dishes well and re-use them.

Avoid kissing or sexual intercourse.

Shower daily if possible but especially for the first 2 days. Rinse the shower or bath well after use.

Wash your clothing and bedclothes separately from other laundry.

If possible, have sole use of a bathroom. Patients (including men) should sit down while urinating. Toilet paper should be used to dry the genitals and should then be flushed down the toilet. Hands should be washed (within the toilet room if possible). Use a separate towel, face cloth, and toothbrush from the rest of the family.

Babies, children, and pregnant women
If you have a baby, it is best for someone else to care for it. If this is not possible, do not have the baby too close to you (i.e., sleeping, sitting on your lap for more than a very short time).

Visits by children and pregnant women should be discouraged. If necessary, try to minimise contact and maximise distance from children and pregnant women.

It is very important to avoid kissing your infant or child for a period of a few weeks as this can transfer radioiodine and result in an unnecessary risk to your child.

Breastfeeding
If you have been breastfeeding your baby, you must stop before radioiodine therapy.

Elderly partners

For people aged 60 years or more, the risk of radiation detriment is small; therefore, only those measures that are easy to take should be encouraged.

Social events

Visits to the cinema and other social events where the patient is in close contact with other individuals for several hours should be avoided.

Returning to work

Patients should not return to work for at least 2 days after treatment. You may return after that if you do not have close contact with other people. If you do have close contact with other people, 1 week may be sufficient, but if you work preparing food for others or work with children or pregnant women, it may be necessary to take several weeks off work. Ask your physician how long you will need to be absent from work.

Emergencies

If you are involved in a traffic accident or another medical emergency, the medical caregivers should be informed of the date, type, and amount of radionuclide therapy you received.

Pregnancy

If you think that you are pregnant and did not know it at the time you received radioiodine, inform your physician immediately. Discuss with your physician how long you should wait before becoming pregnant after radioiodine treatment for hyperthyroidism or thyroid cancer treatment. Typically, pregnancy should be avoided for 4–6 months.

APPENDIX C. SAMPLE CARD FOR PATIENTS GIVEN RADIONUCLIDE THERAPY

Radionuclide
Instruction Card

Radionuclide:	**Iodine-131**
Activity	
	MBq
Administered on ...../......./.......	

Name:
Address:
Hospital No.:
Hospital:
Consultant:

1) Refrain from *all* close contact with children or pregnant women until:

....

2) Refrain from *extended periods* of close contact with children or pregnant women until:

....

3) Avoid prolonged personal contact at home until:

....

4) Avoid prolonged close contact with other people away from home until........................

5) You may return to work on

....

6) Do not sleep with an adult in the same bed until...

Signed:

(Doctor)

This card should be carried at all times until latest date shown on page 2

In case of difficulty, please contact

Telephone

Or the consultant mentioned on page 1

Please contact the hospital if vomiting or incontinence of urine occurs within 24 h of treatment.

REFERENCES

Administration of Radioactive Substances Advisory Committee (2000) UK notes for guidance on the clinical administration of radiopharmaceuticals and use of sealed radioactive sources. Nucl. Med. Commun. 21 (Suppl.), S1–S91.

Aerts, M.G. (2000) Cremation of Corpses Containing Bone-seeking Radionuclides Following Medical Treatment. In: *Proceedings of the IRPA 10 Meeting, International Society of Radiation Protection, Hiroshima, 14–19 May.*

Australian Radiation Protection and Nuclear Safety Agency (2002) Discharge of Patients Undergoing Treatment with Radioactive Substances. Radiation Protection Series No. 4. ARPANSA. http://www.arpansa.gov.au/rps-pubs.htm.

Barrington, S.F., Kettle, A.G., O'Doherty, M.J., et al. (1996a) Radiation dose rates from patients receiving iodine-131 therapy for carcinoma of the thyroid. Eur. J. Nucl. Med. 23, 123–130.

Barrington, S.F., Kettle, A.G., Thompson, W.H., et al. (1996b) RCP guidelines on radiation protection following radioiodine therapy for hyperthyroidism: are they appropriate?. Nucl. Med. Commun. 17, 275.

Barrington, S.F., O'Doherty, M.J., Kettle, A.G., et al. (1999) Radiation exposure of families of outpatients treated with radioactive iodine (iodine-131) for hyperthyroidism. Eur. J. Nucl. Med. 26, 686–692.

Barrington, S.F., Anderson, P., Kettle, A.G., et al. (2003) Does significant contamination occur in families of patients treated with 131-I for hyperthyroidism?. Eur. J. Nucl. Med. 30 (Suppl. 2), S164 (Abstract 69).

Beierwaltes, W.H., Widman, J. (1992) How harmful to others are iodine-131 treated patients: an editorial. J. Nucl. Med. 33, 2116–2117.

British Institute of Radiology (1999). Patients leaving hospital after administration of radioactive substances. Br. J. Radiol. 72, 121–125.

Buchan, R.C.T., Brindle, M.A. (1971) Radioiodine therapy to out-patients – the radiation hazard. Br. J. Radiol. 44, 973–975.

Buchan, R.C.T., Brindle, M.A. (1970) Radioiodine therapy to out-patients – the contamination hazard. Br. J. Radiol. 43, 479–483.

Buettner, C., Surks, M. (2003) Police detainment of a patient following treatment with radioactive iodine. JAMA 288, 2687–2688.

Castronovo, F.P., Beh, R.A., Veilleux, N.M. (1986) Iodine-131 therapy patients: radiation dose to staff. Radiat. Prot. Dosim. 15, 45–49.

Chavin, W., Cukrowski, C.A. (1968) Effects of iodine-131 on the thyroid and pituitary of goldfish (carassius auratus). Radiat. Res. 34, 170–199.

Copplestone, D., Toal, M.E.D., Johnson, M.S., et al. (2000) Environmental effects of radionuclides – observations on natural ecosystems. J. Radiat. Prot. 20, 29–40.

Coover, L.R., Silberstein, E.B., Kuhn, P.J., et al. (2000) Therapeutic ^{131}iodine in outpatients: a simplified method of conforming to the Code of Federal Regulations, Title 10 Part 35.75. J. Nucl. Med. 41, 1868–1875.

Crockett, G. (2000) Sources and fate of discharges of liquid radioactive waste to public sewers. Radiol. Prot. Bull. 226, 19–24.

Culver, C.M., Dworkin, H.J. (1991) Radiation safety considerations for post-iodine-131 hyperthyroid therapy. J. Nucl. Med. 32, 169–173.

deKlerk, J.M.H. (2000) Iodine-131 therapy: inpatient or outpatient?. J. Nucl. Med. 41, 1876–1878.

Denman, A.R., Martin, S. (2001) Care of a terminally ill patient following a thyroid ablation dose of ^{131}I sodium iodide. Br. J. Radiol. 74, 1077–1078.

Dickman, P., Holm, L., Lundell, G., Boice, J., Hall, P. (2003) Thyroid cancer risk after thyroid examination with 131I: a population-based cohort study in Sweden. Int. J. Cancer 106, 580–587.

Driver, I., Packer, S. (2001) Radioactive waste discharge quantities for patients undergoing radioactive iodine therapy for thyroid carcinoma. Nucl. Med. Commun. 22, 1129–1132.

Eisenbud, M. (1973) *Environmental Radioactivity, 2nd edn.* Academic Press, New York.

Erlandsson, B., Mattsson, S. (1978) Medically used radionuclides in sewage sludge. Water Air Soil Pollut. 9, 199–206.
Environment Agency (2000) *UK Investigation of the Sources and Fate of Radioactive Discharges into Public Sewers. R&D Technical Report P288*. Environment Agency, London.
European Commission (1998) *Radiation Protection 97, Radiation Protection Following Iodine-131 (Exposures due to Outpatients or Discharged Inpatients)*. European Commission, Luxembourg.
European Thyroid Association (1996) ^{131}I Therapy for hyperthyroidism towards 2000. Eur. J. Nucl. Med. 23, BP13–BP15.
European Union (1996) Council of the European Union: Guideline 96/29 EURATOM to determine basic directions of health protection of public and workers against ionizing radiation related dangers. Official J. Eur. Community No. L, 159/1.
European Union (1997) Council of the European Union: Council directive 1997/43/ EURATOM on health protection of individuals against ionizing radiation in relation to medical exposure. Official J. Eur. Community No. L, 180.
EURATOM (1997) *Expert Group ex.art 31 Guidance of Radiation Protection Following Iodine Therapy Concerning Doses due to Outpatients or Discharged Inpatients*. EURATOM, Brussels.
Evdokimoff, V., Cash, C., Buckley, K., Cardenas, A. (1994) Potential for radioactive patient excreta in hospital trash and medical waste. Health Phys. 66, 209–211.
Fenner, F.D., Martin, J.E. (1997) Behavior of Na ^{131}I and meta (^{131}I) iodobenzylguanidine (MIBG) in municipal sewerage. Health Phys. 73, 333–339.
Griffiths, P.A., Jones, G.P., Marshall, C., Powley, S.C. (2000) Radiation protection consequences of the care of a terminally ill patient having received a thyroid ablation dose of ^{131}sodium iodide. Br. J. Radiol. 73, 1209–1212.
Gunesekera, R., Thompson, W.H., Harding, L.K. (1996) Use of public transport by ^{131}I therapy outpatients. Nucl. Med. Commun. 17, 275.
Hanscheid, H., Lassmann, M., Reiners, C.H.R., Aulbach, F. (2003) Monitoring of ^{131}I incorporation in nuclear medicine personnel by self accomplished measurements. Nuklearmedizin 42, 45–49.
Harris, P.J. (1959) A study of thyroid function in Fundulus heteroclitus (killifish). Biol. Bull. 117, 91–98.
Hilditch, T.E., Connell, J.M.C., Davies, D.L., et al. (1991) Radiological protection guidance for radioactive patients – new data for therapeutic ^{131}I. Nucl. Med. Commun. 12, 485–495.
Ho, S.Y., Shearer, D.R. (1992) Radioactive contamination in hospitals from nuclear medicine patients. Health Phys. 62.
Homer, L., Smith, A.H. (2002) Radiation protection issues of treating hyperthyroidism with ^{131}I in patients on haemodialysis. Nucl. Med. Commun. 23, 261–264.
Ibis, E., Wilson, C.R., Collier, B.D., et al. (1992) Iodine-131 contamination from thyroid cancer patients. J. Nucl. Med. 33, 2110–2115.
Ilyin, L., Arkhangel'skaya, G.V., Konstantinov, O., Likhtarev, I.A. (1972) *Radioactive Iodine in the Problem of Radiation Safety*. Atomizdat, Moscow.
International Atomic Energy Agency (1996) *Basic Safety Standards*. IAEA,Vienna.
International Atomic Energy Agency (2002a) *Radiological Protection for Medical Exposure to Ionizing Radiation, Safety Guide No. RS-G-1.5*. IAEA, Vienna.
International Atomic Energy Agency (2002b) *Detection of Radioactive Materials at Borders, IAEA-TECDOC-1312*. IAEA, Vienna.
International Commission on Radiological Protection (1987) Radiation dose to patients from radio pharmaceuticals. ICRP Publication 53, AM-ICRP 18(1–4).
International Commission on Radiological Protection (1991) 1990 Recommendations of the International Commision of Radiological Protection. ICRP Publication 60, Ann. ICRP 21 (1–3).
International Commission on Radiological Protection (1996) Radiological protection and safety in medicine. ICRP Publication 73, Ann. ICRP 26 (2).
International Commission on Radiological Protection (1997) Radiological protection policy for the disposal of radioactive waste. ICRP Publication 77, Ann. ICRP 27 (Suppl.).
International Commission on Radiological Protection (1999) ICRP Database of Dose Coefficients. Workers and Members of the Public, CD-ROM Distributed by Elsevier Ltd, Oxford.

International Commission on Radiological Protection (2000) Pregnancy and medical radiation. ICRP Publication 84, Ann ICRP 30 (1).
International Commission on Radiological Protection (2002) Doses to the embryo and fetus from intakes of radionuclides by the mother. ICRP Publication 88, Ann. ICRP 31 (1–3).
International Commission on Radiological Protection (2003) A framework for assessing the impact of ionising radiation on non-human species. ICRP Publication 91, Ann. ICRP 33(3).
Jacobson, A.P., Plato, P.A., Toeroek, D. (1978) Contamination of the home environment by patients treated with iodine-131: initial results. Am. J. Public Health 68, 225–230.
Johnson, T.K., Lin, E., Haugen, B. (2002) Contamination as a route for ^{131}I uptake in family members: implications for public health. J. Nucl. Med. 43 (Suppl.), abstract.
Johnston, A.S., Minarcik, J., Rossi, R., et al. (1979) Autopsy experience with a radioactive cadaver. Health Phys. 37, 231–236.
Kelly, J., Finch, H. (2002) Benchmarking Public Opinion on the Management of Radioactive Waste. *A Report for the Radioactive Substances Division of the Department for the Environment, Food and Rural Affairs.*
Kerber, R., Till, J., Simon, S., et al. (1993) A cohort study of thyroid disease in relation to fallout from nuclear weapons testing. JAMA 270, 2076.
Larsen, P.R., Conard, R. (1978) Thyroid hypofunction appearing as a delayed manifestation of accidental exposure to radioactive fallout in a Marshallese population. Brookhaven National Laboratory Report BNL-24104.
LaRoche, G., LeBlond, C.P. (1954) Destruction of the thyroid gland of Atlantic salmon by means of radioiodine. Proc. Soc. Exp. Biol. Med. 87, 273–276.
LaRoche, G., Johnson, C.L., Woodall, A.N. (1965) Thyroid function in the rainbow trout: biological and histological evidence of radiothyroidectomy. Gen. Comp. Endocrinol. 5, 145–159.
Lassmann, M., Hanscheid, H., Schelper, L.F. (1998) Measurement of incorporation in family members of patients with benign thyroid disease after radioiodine therapy. Nuklearmedizin 37, 120–123.
Leslie, W.D., Havelock, J., Palser, R., Abrams, D.N. (2002) Large-body radiation doses following radioiodine therapy. Nucl. Med. Commun. 23, 1091–1097.
Leung, P.M., Nikolic, M. (1998) Disposal of therapeutic iodine-131 waste using a multiple holding tank system. Health Phys. 75, 315–321.
Lubin, E. (2002) Definitive improvement in the approach to the treated patient as a radioactive source. J. Nucl. Med. 43, 364–365.
Martin, J.E., Fenner, F.D. (1997) Radioactivity in municipal sewerage and sludge. Public Health Rep., 308–316.
Marcus, C.S., Aldrich, R. (1997) Avoiding solid waste contamination problems from iodine-131 patients. J. Nucl. Med. 38, 26N.
Mathieu, I., Caussin, J., Smeesters, P., Wamberseie, A., Beckers, C. (1997) Doses in family members after ^{131}I treatment. Lancet 350, 1074–1075.
Mathieu, I., Caussin, J., Smeesters, P., et al. (1999) Recommended restrictions after iodine-131 therapy: measured doses in family members. Health Phys. 76, 129–136.
Meck, R.A. (1996) Complete decay of radionuclides: implications for low-level waste disposal in municipal landfills. Health Phys. 70, 706–7111.
Monsieurs, M., Thierens, H., Dierckx, R.A., et al. (1998) Real-life radiation burden to relatives of patients treated with iodine-131: a study in 8 centers in Flanders (Belgium). Eur. J. Nucl. Med. 25, 1368–1376.
Moreno Garcia, L. (2001) Expectations of Patients' Advocates, Radiological Protection of Patients in Diagnostic and Interventional Radiology, Nuclear Medicine and Radiotherapy. In: *Proceedings of an International Conference, Malaga, 26–30 March 2001.* International Atomic Energy Agency, Vienna.
National Council on Radiation Protection and Measurements (1970) *Precautions in the Management of Patients who have Received Therapeutic Amounts of Radionuclides. Report 37.* NCRP, Bethesda.
National Council on Radiation Protection and Measurements (1995) Dose Limits for Individuals who Receive Exposure from Radionuclide Therapy Patients. NCRP Commentary No. 11. NCRP, Bethesda.
National Health and Medical Research Council (1987) Australian Code of Practice for the Safe Handling of Corpses Containing Radioactive Materials (1986). Australian Government Publishing Service, Canberra.

National Radiation Protection Board (1988) Guidance Notes for the Protection of Persons Against Ionising Radiations Arising from Medical and Dental Use. NRPB, Oxon.
National Radiation Protection Board (2000) Institute of Physics and Engineering in Medicine, National Radiological Protection Board, Health and Safety Executive, The Health Departments 2000 Medical and Dental Guidance Notes – a Good Practice Guide to Implement Ionising Radiation Protection Legislation in the Clinical Environment.
Norris, D.O., Gorbman, A. (1965) Radiothyroidectomy of larval steelhead trout. Proc. Soc. Exp. Biol. Med. 119, 1205–1207.
Nishizawa, K., Ohara, K., Ohshima, M., et al. (1980) Monitoring of I excretions and used materials of patients treated with ^{131}I. Health Phys. 38, 467–481.
Nuclear Energy Agency (1995) The Environmental and Ethical Basis of Geological Disposal of Long-lived Radioactive Wastes. A Collective Opinion of the Radioactive Waste Management Committee of the OECD Nuclear Energy Agency. NEA, Paris.
O'Doherty, M.J., Kettle, A.G., Eustance, C.N.P., et al. (1993) Radiation dose rates from adult patients receiving ^{131}I therapy for hyperthyroidism. Nucl. Med. Commun. 14, 160–168.
Olivereau, M., LaRoche, G. (1965) Effects of low iodine intake or destructive doses of iodine-131 on pituitary histochemistry of young Chinook salmon. Am. Zool. 5, 234.
Olivereau, M. (1957) Radiothyroidectomie chez l'Anguille. Arch. Anat. Microscop. Morphol. Exptl. 46, 39–59.
Osborn, S., Greaves, C., Tindale, W. (2002) Care of a helpless patient and handling of the radioactive corpse. J. Radiol. Protect. 22, 185–187.
Parthasarathy, K.L., et al. (1982) Necropsy of a cadaver containing 50 mCi of sodium 131 iodide. J. Nucl. Med. 23, 777–780.
Pentreath, R.J. (2002) Radiation protection of people and the environment: developing a common approach. J. Radiol. Protect. 22, 45–46.
Prichard, H.M., Gesell, T.F., Davis, E. (1981) Iodine-131 levels in sludge and treated municipal wastewaters near a large medical complex. Am. J. Public Health 71, 47–52.
Rallison, M.L., Dobyns, B.M., Keating, F.R., et al. (1974) Thyroid disease in children: a survey of subjects potentially exposed to fallout radiation. Am. J. Med. 56, 457–463.
Reiners, C., Lassmann, M. (1999) Radioiodine (^{131}I) treatment of hyperthyroidism: radiation protection and quality assurance. Eur. J. Nucl. Med. 26, 683–685.
Ron, E., Lubin, J.H., Shore, R.E., et al. (1995) Thyroid cancer after exposure to external radiation: a pooled analysis of seven studies. Radiat. Res. 141, 259–277.
Rutar, F.J., Augustine, S.C., Colcher, D., et al. (2001) Outpatient treatment with iodine-131 anti-B1 antibody: radiation exposure to family members. J. Nucl. Med. 42, 90–915.
Saenger, E.L., Kereiakes, J.G. (1980) Reflections on cancer treatment and the federal agency regulations. Radiology 137, 865–866.
Schomaecker, K., Fischer, T., Gaidouk, M., et al. (2000) Exhalation of iodine-131 after radioiodine therapy: time dependence and chemical form. J. Nucl. Med. 41 (Suppl.), 251P.
Shishkanov, N.G., Bakun, Y.M., Roziev, R.A., et al. (2001) Radiation protection of members of the public contacting patients after therapeutic treatment with radioiodine. Med. Radiol. Radiat. Protect. 5, 34–46 (in Russian).
Siegel, J.A. (1999) Outpatient Radionuclide Therapy. In: *Proceedings of the Thirty-Fifth Annual Meeting of the National Council on Radiation Protection and Measurements, Proceedings No. 21, 7–8 April 1999, Arlington VA*. NCRP, Bethesda.
Siegel, J.A., Marcus, C.S., Sparks, R.B. (2002a) Calculating the absorbed dose from radioactive patients: the line-source versus point-source model. J. Nucl. Med. 43, 1241–1244.
Siegel, J.A., Kroll, S., Regan, D., et al. (2002b) A practical methodology for patient release after tositumomab and iodine-131 tositumomab therapy. J. Nucl. Med. 43, 354–363.
Siegel, J.A., Sparks, R.B. (2002) Radioactivity appearing at landfills in household trash of nuclear medicine patients: much ado about nothing? Health Phys. 82, 367–372.
Silberstein, E.B., Taylor, A.T. Jr. (1996) Society of Nuclear Medicine Guideline for Bone Pain Treatment. J. Nucl. Med. 37, 881–884.

Silberstein, E.B., Buscombe, J.R., Taylor Jr., A.T. (2003) Society of Nuclear Medicine Procedure Guideline for Palliative Treatment of Painful Bone Metastases. Version 3.0. Available at www.snm.org.

Sodd, V.J., Velten, R.J., Saenger, E.L. (1975) Concentrations of the medically useful radionuclides, technetium-99m and iodine-131 at a large metropolitan waste water treatment plant. Health Phys. 28, 355–359.

Swedish Radiation Protection Institute (2000) Regulations and General Advice on Nuclear Medicine. SSI FS 2000-3.

Thompson, W., Williams, N.R., Harding, L.K. (1994) A model of excreted activities of radiopharmaceuticals from patients to the drains. Eur. J. Nucl. Med. 21, 876.

UK Department of the Environment (2002) UK Strategy for Radioactive Discharges 2001–2020. Department for Environment, Food and Rural Affairs, London.

United Nations Scientific Committee on the Effects of Atomic Radiation (2000) Sources and Effects of Ionizing Radiation. *2000 Report to the General Assembly with Annexes, United Nations, Vienna.*

US Nuclear Regulatory Commission (1997a) Regulatory Analysis on Criteria for the Release of Patients Administered Radioactive Material, NUREG-1492. USNRC, Washington.

US Nuclear Regulatory Commission (1997b) Release of Patients Administered Radioactive Materials. Regulatory Guide 8.39. USNRC, Washington.

Wellner, U., Eschner, W., Hillger, W., et al. (1998) The exposure of relatives to patients of a nuclear medicine ward after radioiodine therapy by inhalation of ^{131}I in their home. Nuklearmedizin 37, 113–119.

FSC
www.fsc.org
MIX
Papier | Fördert
gute Waldnutzung
FSC® C083411

Kolibri 360 GmbH
Ferdinand-Jühlke-Straße 7
99095 Erfurt, Deutschland
produktsicherheit@kolibri360.de